Pitman Research Notes in Mathematics Series

Submission of proposals for consideration
Suggestions for publication, in the form of outlines and representative samples, are invited by the Editorial Board for assessment. Intending authors should approach one of the main editors or another member of the Editorial Board, citing the relevant AMS subject classifications. Alternatively, outlines may be sent directly to the publisher's offices. Refereeing is by members of the board and other mathematical authorities in the topic concerned, throughout the world.

Preparation of accepted manuscripts
On acceptance of a proposal, the publisher will supply full instructions for the preparation of manuscripts in a form suitable for direct photo-lithographic reproduction. Specially printed grid sheets are provided and a contribution is offered by the publisher towards the cost of typing. Word processor output, subject to the publisher's approval, is also acceptable.

Illustrations should be prepared by the authors, ready for direct reproduction without further improvement. The use of hand-drawn symbols should be avoided wherever possible, in order to maintain maximum clarity of the text.

The publisher will be pleased to give any guidance necessary during the preparation of a typescript, and will be happy to answer any queries.

Important note
In order to avoid later retyping, intending authors are strongly urged not to begin final preparation of a typescript before receiving the publisher's guidelines and special paper. In this way it is hoped to preserve the uniform appearance of the series.

Longman Scientific & Technical
Longman House
Burnt Mill
Harlow, Essex, UK
(tel (0279) 26721)

Titles in this series

52 Wave propagation in viscoelastic media
F Mainardi
53 Nonlinear partial differential equations and their applications: Collège de France Seminar. Volume I
H Brezis and J L Lions
54 Geometry of Coxeter groups
H Hiller
55 Cusps of Gauss mappings
T Banchoff, T Gaffney and C McCrory
56 An approach to algebraic K-theory
A J Berrick
57 Convex analysis and optimization
J-P Aubin and R B Vintner
58 Convex analysis with applications in the differentiation of convex functions
J R Giles
59 Weak and variational methods for moving boundary problems
C M Elliott and J R Ockendon
60 Nonlinear partial differential equations and their applications: Collège de France Seminar. Volume II
H Brezis and J L Lions
61 Singular systems of differential equations II
S L Campbell
62 Rates of convergence in the central limit theorem
Peter Hall
63 Solution of differential equations by means of one-parameter groups
J M Hill
64 Hankel operators on Hilbert space
S C Power
65 Schrödinger-type operators with continuous spectra
M S P Eastham and H Kalf
66 Recent applications of generalized inverses
S L Campbell
67 Riesz and Fredholm theory in Banach algebra
B A Barnes, G J Murphy, M R F Smyth and T T West
68 Evolution equations and their applications
F Kappel and W Schappacher
69 Generalized solutions of Hamilton-Jacobi equations
P L Lions
70 Nonlinear partial differential equations and their applications: Collège de France Seminar. Volume III
H Brezis and J L Lions
71 Spectral theory and wave operators for the Schrödinger equation
A M Berthier
72 Approximation of Hilbert space operators I
D A Herrero
73 Vector valued Nevanlinna Theory
H J W Ziegler
74 Instability, nonexistence and weighted energy methods in fluid dynamics and related theories
B Straughan
75 Local bifurcation and symmetry
A Vanderbauwhede
76 Clifford analysis
F Brackx, R Delanghe and F Sommen
77 Nonlinear equivalence, reduction of PDEs to ODEs and fast convergent numerical methods
E E Rosinger
78 Free boundary problems, theory and applications. Volume I
A Fasano and M Primicerio
79 Free boundary problems, theory and applications. Volume II
A Fasano and M Primicerio
80 Symplectic geometry
A Crumeyrolle and J Grifone
81 An algorithmic analysis of a communication model with retransmission of flawed messages
D M Lucantoni
82 Geometric games and their applications
W H Ruckle
83 Additive groups of rings
S Feigelstock
84 Nonlinear partial differential equations and their applications: Collège de France Seminar. Volume IV
H Brezis and J L Lions
85 Multiplicative functionals on topological algebras
T Husain
86 Hamilton-Jacobi equations in Hilbert spaces
V Barbu and G Da Prato
87 Harmonic maps with symmetry, harmonic morphisms and deformations of metrics
P Baird
88 Similarity solutions of nonlinear partial differential equations
L Dresner
89 Contributions to nonlinear partial differential equations
C Bardos, A Damlamian, J I Díaz and J Hernández
90 Banach and Hilbert spaces of vector-valued functions
J Burbea and P Masani
91 Control and observation of neutral systems
D Salamon
92 Banach bundles, Banach modules and automorphisms of C*-algebras
M J Dupré and R M Gillette
93 Nonlinear partial differential equations and their applications: Collège de France Seminar. Volume V
H Brezis and J L Lions
94 Computer algebra in applied mathematics: an introduction to MACSYMA
R H Rand
95 Advances in nonlinear waves. Volume I
L Debnath
96 FC-groups
M J Tomkinson
97 Topics in relaxation and ellipsoidal methods
M Akgül
98 Analogue of the group algebra for topological semigroups
H Dzinotyiweyi
99 Stochastic functional differential equations
S E A Mohammed

100 Optimal control of variational inequalities
V Barbu
101 Partial differential equations and dynamical systems
W E Fitzgibbon III
102 Approximation of Hilbert space operators. Volume II
C Apostol, L A Fialkow, D A Herrero and D Voiculescu
103 Nondiscrete induction and iterative processes
V Ptak and F-A Potra
104 Analytic functions – growth aspects
O P Juneja and G P Kapoor
105 Theory of Tikhonov regularization for Fredholm equations of the first kind
C W Groetsch
106 Nonlinear partial differential equations and free boundaries. Volume I
J I Díaz
107 Tight and taut immersions of manifolds
T E Cecil and P J Ryan
108 A layering method for viscous, incompressible L_p flows occupying R^n
A Douglis and E B Fabes
109 Nonlinear partial differential equations and their applications: Collège de France Seminar. Volume VI
H Brezis and J L Lions
110 Finite generalized quadrangles
S E Payne and J A Thas
111 Advances in nonlinear waves. Volume II
L Debnath
112 Topics in several complex variables
E Ramírez de Arellano and D Sundararaman
113 Differential equations, flow invariance and applications
N H Pavel
114 Geometrical combinatorics
F C Holroyd and R J Wilson
115 Generators of strongly continuous semigroups
J A van Casteren
116 Growth of algebras and Gelfand–Kirillov dimension
G R Krause and T H Lenagan
117 Theory of bases and cones
P K Kamthan and M Gupta
118 Linear groups and permutations
A R Camina and E A Whelan
119 General Wiener–Hopf factorization methods
F-O Speck
120 Free boundary problems: applications and theory, Volume III
A Bossavit, A Damlamian and M Fremond
121 Free boundary problems: applications and theory, Volume IV
A Bossavit, A Damlamian and M Fremond
122 Nonlinear partial differential equations and their applications: Collège de France Seminar. Volume VII
H Brezis and J L Lions
123 Geometric methods in operator algebras
H Araki and E G Effros
124 Infinite dimensional analysis–stochastic processes
S Albeverio
125 Ennio de Giorgi Colloquium
P Krée
126 Almost-periodic functions in abstract spaces
S Zaidman
127 Nonlinear variational problems
A Marino, L Modica, S Spagnolo and M Degiovanni
128 Second-order systems of partial differential equations in the plane
L K Hua, W Lin and C-Q Wu
129 Asymptotics of high-order ordinary differential equations
R B Paris and A D Wood
130 Stochastic differential equations
R Wu
131 Differential geometry
L A Cordero
132 Nonlinear differential equations
J K Hale and P Martinez-Amores
133 Approximation theory and applications
S P Singh
134 Near-rings and their links with groups
J D P Meldrum
135 Estimating eigenvalues with *a posteriori/a priori* inequalities
J R Kuttler and V G Sigillito
136 Regular semigroups as extensions
F J Pastijn and M Petrich
137 Representations of rank one Lie groups
D H Collingwood
138 Fractional calculus
G F Roach and A C McBride
139 Hamilton's principle in continuum mechanics
A Bedford
140 Numerical analysis
D F Griffiths and G A Watson
141 Semigroups, theory and applications. Volume I
H Brezis, M G Crandall and F Kappel
142 Distribution theorems of L-functions
D Joyner
143 Recent developments in structured continua
D De Kee and P Kaloni
144 Functional analysis and two-point differential operators
J Locker
145 Numerical methods for partial differential equations
S I Hariharan and T H Moulden
146 Completely bounded maps and dilations
V I Paulsen
147 Harmonic analysis on the Heisenberg nilpotent Lie group
W Schempp
148 Contributions to modern calculus of variations
L Cesari
149 Nonlinear parabolic equations: qualitative properties of solutions
L Boccardo and A Tesei
150 From local times to global geometry, control and physics
K D Elworthy

151 A stochastic maximum principle for optimal control of diffusions
U G Haussmann
152 Semigroups, theory and applications. Volume II
H Brezis, M G Crandall and F Kappel
153 A general theory of integration in function spaces
P Muldowney
154 Oakland Conference on partial differential equations and applied mathematics
L R Bragg and J W Dettman
155 Contributions to nonlinear partial differential equations. Volume II
J I Díaz and P L Lions
156 Semigroups of linear operators: an introduction
A C McBride
157 Ordinary and partial differential equations
B D Sleeman and R J Jarvis
158 Hyperbolic equations
F Colombini and M K V Murthy
159 Linear topologies on a ring: an overview
J S Golan
160 Dynamical systems and bifurcation theory
M I Camacho, M J Pacifico and F Takens
161 Branched coverings and algebraic functions
M Namba
162 Perturbation bounds for matrix eigenvalues
R Bhatia
163 Defect minimization in operator equations: theory and applications
R Reemtsen
164 Multidimensional Brownian excursions and potential theory
K Burdzy
165 Viscosity solutions and optimal control
R J Elliott
166 Nonlinear partial differential equations and their applications. Collège de France Seminar. Volume VIII
H Brezis and J L Lions
167 Theory and applications of inverse problems
H Haario
168 Energy stability and convection
G P Galdi and B Straughan
169 Additive groups of rings. Volume II
S Feigelstock
170 Numerical analysis 1987
D F Griffiths and G A Watson
171 Surveys of some recent results in operator theory. Volume I
J B Conway and B B Morrel
172 Amenable Banach algebras
J-P Pier
173 Pseudo-orbits of contact forms
A Bahri
174 Poisson algebras and Poisson manifolds
K H Bhaskara and K Viswanath
175 Maximum principles and eigenvalue problems in partial differential equations
P W Schaefer
176 Mathematical analysis of nonlinear, dynamic processes
K U Grusa
177 Cordes' two-parameter spectral representation theory
D F McGhee and R H Picard
178 Equivariant K-theory for proper actions
N C Phillips
179 Elliptic operators, topology and asymptotic methods
J Roe
180 Nonlinear evolution equations
J K Engelbrecht, V E Fridman and E N Pelinovski
181 Nonlinear partial differential equations and their applications. Collège de France Seminar. Volume IX
H Brezis and J L Lions
182 Critical points at infinity in some variational problems
A Bahri
183 Recent developments in hyperbolic equations
L Cattabriga, F Colombini, M K V Murthy and S Spagnolo
184 Optimization and identification of systems governed by evolution equations on Banach space
N U Ahmed
185 Free boundary problems: theory and applications. Volume I
K H Hoffmann and J Sprekels
186 Free boundary problems: theory and applications. Volume II
K H Hoffmann and J Sprekels
187 An introduction to intersection homology theory
F Kirwan
188 Derivatives, nuclei and dimensions on the frame of torsion theories
J S Golan and H Simmons
189 Theory of reproducing kernels and its applications
S Saitoh
190 Volterra integrodifferential equations in Banach spaces and applications
G Da Prato and M Iannelli
191 Nest algebras
K R Davidson
192 Surveys of some recent results in operator theory. Volume II
J B Conway and B B Morrel
193 Nonlinear variational problems. Volume II
A Marino and M K Murthy
194 Stochastic processes with multidimensional parameter
M E Dozzi
195 Prestressed bodies
D Iesan
196 Hilbert space approach to some classical transforms
R H Picard
197 Stochastic calculus in application
J R Norris
198 Radical theory
B J Gardner
199 The C* – algebras of a class of solvable Lie groups
X Wang

200 Stochastic analysis, path integration and dynamics
D Elworthy
201 Riemannian geometry and holonomy groups
S Salamon
202 Strong asymptotics for extremal errors and polynomials associated with Erdös type weights
D S Lubinsky
203 Optimal control of diffusion processes
V S Borkar
204 Rings, modules and radicals
B J Gardner
205 Numerical studies for nonlinear Schrödinger equations
B M Herbst and J A C Weideman
206 Distributions and analytic functions
R D Carmichael and D Mitrović
207 Semicontinuity, relaxation and integral representation in the calculus of variations
G Buttazzo
208 Recent advances in nonlinear elliptic and parabolic problems
P Bénilan, M Chipot, L Evans and M Pierre
209 Model completions, ring representations and the topology of the Pierce sheaf
A Carson
210 Retarded dynamical systems
G Stepan
211 Function spaces, differential operators and nonlinear analysis
L Paivarinta
212 Analytic function theory of one complex variable
C C Yang, Y Komatu and K Niino
213 Elements of stability of visco-elastic fluids
J Dunwoody
214 Jordan decompositions of generalised vector measures
K D Schmidt
215 A mathematical analysis of bending of plates with transverse shear deformation
C Constanda
216 Ordinary and partial differential equations Vol II
B D Sleeman and R J Jarvis
217 Hilbert modules over function algebras
R G Douglas and V I Paulsen
218 Graph colourings
R Wilson and R Nelson
219 Hardy-type inequalities
A Kufner and B Opic
220 Nonlinear partial differential equations and their applications. College de France Seminar Volume X
H Brezis and J L Lions

Ordinary and partial differential equations, Volume II

B D Sleeman & R J Jarvis (Editors)

University of Dundee

Ordinary and partial differential equations, Volume II

Proceedings of the tenth Dundee Conference, 1988

Copublished in the United States with
John Wiley & Sons, Inc., New York

Longman Scientific & Technical,
Longman Group UK Limited,
Longman House, Burnt Mill, Harlow
Essex CM20 2JE, England
and Associated Companies throughout the world.

Copublished in the United States with
John Wiley & Sons, Inc., 605 Third Avenue, New York, NY 10158

First published 1989

AMS Subject Classification: (Main) 34–06, 35–06
(Subsidiary) 78–06, 92–06

ISSN 0269-3674

British Library Cataloguing in Publication Data
Ordinary and partial differential equations,
proceedings of the tenth Dundee
Conference, 1988.
1. Ordinary differential equations 2.
Partial differential equations
I. Sleeman, B.D. (Brian D) II. Jarvis,
R.J. (Richard James)
515.3′52
ISBN 0-582-03188-5

Library of Congress Cataloging-in-Publication Data
Data applied for

Printed and bound in Great Britain
by Biddles Ltd, Guildford and King's Lynn

Contents

Preface

These Proceedings form a record of the lectures delivered at the Tenth Conference on Ordinary and Partial Differential Equations, which was held at the University of Dundee, Scotland during July 1988.

The Conference was attended by more than 100 mathematicians from Australia, Europe, North America, the People's Repulic of China, Egypt, Iran, Iraq, Pakistan, Poland and the USSR.

It was organised by us and we thank all the mathematicians who took part in the work of the Conference.

We also thank the University of Dundee for generously supporting the Conference; the Warden and Staff of West Park Hall for their help in providing accommodation for the Participants; Colleagues for their assistance during the week of the Conference, and the Bursar of Residences and the Finance Officer of the University of Dundee.

It is a pleasure to record special appreciation of grants from the Science and Engineering Research Council and the European Research Office of the United States Army. These made available travel funds for the Plenary Speakers and also helped to provide secretarial services for the Conference.

Special thanks are due to Miss Shirley Peter, Secretary in the Department of Mathematics and Computer Science of the University of Dundee, for her invaluable contribution to the work and organisation of the Conference.

B.D. Sleeman R.J. Jarvis

List of Contributors

A. Bayliss	The Technological Institute, Northwestern University, Evanston, Illinois, U.S.A.
P.J. Browne	Calgary University, Calgary, Alberta, Canada.
J. Fleckinger-Pellé	Université Paul Sabatier, Toulouse, France.
R. Leis	Universität Bonn, Bonn, West Germany.
H.A. Levine	Iowa State University, Ames, Iowa, U.S.A.
B.J. Matkowsky	The Technological Institute, Northwestern University, Evanston, Illinois, U.S.A.
J. Mawhin	Université Catholique de Louvain, Louvain-la-Neuve, Belgium.
E. Meister	Technische Hochschule Darmstadt, Darmstadt, West Germany.
M. Minkoff	Argonne National Laboratory, Argonne, Illinois, U.S.A.
R.A. Smith	University of Durham, Durham, U.K.
F.-O. Speck	Technische Hochschule Darmstadt, Darmstadt, West Germany.
Tian Jinghuang	Chinese Academy of Sciences, Chengdu, China.
J.F. Toland	University of Bath, Bath, U.K.
J. Tyson	Virginia Polytechnic Institute and State University, Blacksburg, Virginia, U.S.A.

A. BAYLISS, B.J. MATKOWSKY AND M. MINKOFF

Bifurcation and pattern formation in combustion

1. INTRODUCTION

In our research program we employ a combination of analytical and numerical methods to determine the behavior of solutions of combustion problems. Specifically we consider highly nonlinear time-dependent systems of partial differential equations which model the behavior of both solid and gaseous fuel combustion. In gaseous fuel combustion we are particularly interested in the transition from laminar to turbulent combustion, including a description of the intermediate stages of this transition. These stages often occur as a sequence of bifurcations, as critical parameters of the problem are varied, with each successive step exhibiting more and more complex spatial and temporal behavior, often leading to spatial and temporal pattern formation. The solutions frequently exhibit very steep gradients, in both time and space, thus naturally calling for adaptive gridding techniques. We have developed an adaptive pseudo-spectral method which is both very accurate and very efficient. Our algorithm allows us to describe the solution on bifurcation branches, well beyond the region where analytical methods work well. We have however taken advantage of the analytical results that we first obtain, to aid us in choosing appropriate parameter values and initial conditions for the numerical computations. In addition the analytical results serve as benchmarks for our computations. The computations reveal new and interesting behavior, not otherwise obtainable. Below we discuss two problems, involving solid and gaseous fuel combustion respectively.

2. CONDENSED-PHASE COMBUSTION

We first consider a problem in gasless condensed-phase combustion. This type of combustion is characterized by a highly exothermic reaction occurring in

* This research was supported in part by the Applied Mathematical Sciences subprogram of the Office of Energy Research, U.S. Department of Energy, under Contract W-31-109-Eng-38 and Grant DEFG02-87ER25027, and N.S.F. Grant DMS87-01543.

the solid fuel itself without the prior formation of a gaseous phase. Thus the solid itself burns, and is transformed directly into a solid product. Owing to the exothermic reaction a combustion wave propagates from the high-temperature combustion products toward the cold unburned fuel. Typically the activation energies of the reaction are large and the reaction is significant only in a narrow region, called the reaction zone, whose width is inversely proportional to the activation energy. In the limit of infinite activation energy the reaction zone shrinks to a propagating surface, termed a reaction front. It has been observed that this process is often accompanied by the melting of the reactant [5], so that a melting front propagates ahead of the reaction zone or front. Upon reacting, the heat of fusion is released, and the product is in the solid phase.

This process is currently being employed as a method of effectively synthesizing certain ceramic and metallic alloys. Rather than employing an external source of energy, the process, referred to as SHS (for self-propagating, high-temperature synthesis), employs the energy of the reaction to convert reactants to products which are specially hard, are impervious to extreme temperatures and have other desired characteristics [6, 14]. It has been observed that the product is not always uniform in composition, but rather there are zones of varying concentrations in the synthesized sample. This is due to pulsations in the velocity of the reaction front. Planar pulsating modes of propagation have been observed, which are sometimes referred to as "auto-oscillatory" combustion, characterized by striations in the synthesized product. Helical modes of propagation, referred to as spin combustion, have also been observed, in which a spiraling motion of a non-uniform front occurs (one or more luminous points, corresponding to hotspots, are observed to move in a helical fashion on the surface of a cylindrical sample) [9, 15].

Mathematically this process can be described by a reaction-diffusion system for the temperature and concentration of a limiting component of the reaction. The reaction is modeled by first-order irreversible Arrhenius kinetics. Typically the mass diffusivity is assumed to be zero in both the solid and liquid phase.

Analysis of a model in which melting was not accounted for, in the limit of infinite activation energy, revealed that auto-oscillatory combustion was due to a Hopf bifurcation when a parameter λ exceeded a critical value λ_c [13].

The bifurcation parameter is $\lambda = N(1 - \sigma)/2$, where N is an appropriately nondimensionalized activation energy and $\sigma = T_u/T_b$, where T_u (T_b) is the temperature of the unburned (burned) solid. This model was extended in [10] to account for melting of the fuel prior to the reaction. Again a similar Hopf bifurcation was found when a parameter $\mu = \lambda/(1-M)$ exceeded a critical value μ_c. The parameter M accounts for the effect of melting and is defined below. In [17], the model, without melting, in a cylindrical geometry, was suggested as a description of spin combustion. However, in the absence of melting, the resulting bifurcation is subcritical and unstable and therefore may not be able to account for the phenomenon. In [11] it was found that spin combustion could be explained as a supercritical, stable Hopf bifurcation. These analytical studies were based on the limit $N \to \infty$, where the reaction zone shrinks to a reaction front and the reaction term becomes asymptotically a δ function on the front, whose strength, as a function of T and of μ, was determined by the method of matched asymptotic expansions.

Bifurcation analysis is necessarily a local theory, valid only in a neighborhood of the bifurcation point. The behavior of solutions far from this neighborhood must be obtained numerically. A related model of gasless condensed-phase combustion was studied numerically in [16]. Sinusoidal oscillations were computed, which took on the character of relaxation oscillations as the activation energy increased. As the activation energy increased further an additional spike in the temperature was observed and the solution appeared to have doubled in period. Upon further increasing the activation energy, additional spikes in the temperature were computed and the pulsation became increasingly complex. In [1], another related model was studied. Again a sinusoidal oscillation was found beyond a critical value of a parameter related to the activation energy. As the parameter was increased further, the authors exhibited oscillations with complex structure, and claimed to have found two transitions, in each of which the period approximately doubled, before the computations had to be stopped due to computational difficulties.

In [3] the model of [10] for finite N was studied numerically. A sinusoidal oscillation in the solution was found very close to the analytically predicted Hopf bifurcation point. There followed a progressive sharpening of the peaks leading to relaxation oscillations. A period-doubling transition was found, and evidence clearly indicated that the transition was due to a period-doubling secondary bifurcation. This model was studied further in [19]. In view of the results in [1, 16] it might have been expected that additional period-doubling bifurcations would be found, with a possible transition to chaos. However, we showed that this is not the case for the parameter range studied. Our results are illustrated in figure 1a and can be summarized as follows: There is a very rapid growth and sharpening of the pulsation along the period-doubled solution branch. Beyond a certain value of μ, stable period-doubled solutions can no longer be computed. There is an interval of bistability in which singly and doubly periodic solutions are both stable, each with its own domain of attraction. No additional period doublings were found along the period-doubled solution branch for the range of parameter values considered, though they may occur for other parameter values.

We solve a model which is a generalization of the one employed in [13] in that it accounts for melting [10]. The reaction term is governed by global, one-step, irreversible Arrhenius kinetics, which is cut off at a certain distance ahead of the melting front.

To describe the model we let a tilde (~) stand for a dimensional quantity, assume the front propagates in the $-\tilde{x}$ direction and denote the location of the melting front by $\tilde{x} = \tilde{\phi}(\tilde{t})$. If $\tilde{T}$ and $\tilde{C}$ respectively denote the temperature and concentration of a limiting component of the reactant, the model is described by the reaction-diffusion system

$$\begin{aligned} \tilde{T}_{\tilde{t}} &= \tilde{\lambda}\tilde{T}_{\tilde{x}\tilde{x}} + \begin{pmatrix} \tilde{\beta} \\ \alpha(\tilde{\beta} + \tilde{\gamma}) \end{pmatrix} g\tilde{A}\tilde{C}\exp(-\tilde{E}/R\tilde{T}) \\ \tilde{C}_{\tilde{t}} &= -\begin{pmatrix} 1 \\ \alpha \end{pmatrix} g\tilde{A}\tilde{C}\exp(-\tilde{E}/R\tilde{T}), \end{aligned} \tag{2.1}$$

where

$$\begin{pmatrix} a \\ b \end{pmatrix} = \begin{matrix} a, & \tilde{x} < \tilde{\phi}(\tilde{t}) \\ b, & \tilde{x} > \tilde{\phi}(\tilde{t}). \end{matrix}$$

In (2.1) $\tilde{\lambda}$ is the thermal conductivity, $\tilde{A}$ the rate constant, $\tilde{\beta}$ the heat of reaction, $\tilde{E}$ the activation energy, and R the gas constant. Because the fuel melts, the rate constant is multiplied by a factor $\alpha > 1$, due to the increased surface-to-surface contact in the liquid phase. Upon melting, the heat of fusion $\tilde{\gamma}$ is absorbed by the fuel, but is released during the reaction so the product is in the solid phase. Thus behind the melting front we take the heat released to be $\tilde{\beta} + \tilde{\gamma}$. The function $g = g(\tilde{x} - \tilde{\phi}(\tilde{t}))$ cuts off the reaction term at some point ahead of the melting front. This is employed because the Arrhenius model for the reaction term does not vanish far ahead of the front, while in practice no significant reaction occurs prior to melting. In the computations we use

$$g(\tilde{x} - \tilde{\phi}(\tilde{t})) = \begin{cases} 1, & \tilde{x} - \tilde{\phi}(\tilde{t}) > Z_c \\ 0, & \tilde{x} - \tilde{\phi}(\tilde{t}) < Z_c \end{cases},$$

where $Z_c = -3$. We have found that the behavior of the solution is not sensitive to Z_c in this range although the location of the period-doubled bifurcation point varies slightly with Z_c [3]. In addition, no significant effect is found if a smoother functional form is used for the function g.

Across the melting front there is a jump in the heat flux, due to the absorption of the heat of fusion necessary to cause melting. The velocity $\tilde{\phi}_{\tilde{t}}$ of the melting front satisfies

$$\tilde{\phi}_{\tilde{t}} = \frac{-\tilde{\lambda}}{\tilde{\gamma}\tilde{C}_m} [\, \tilde{T}_{\tilde{x}} \,], \tag{2.2}$$

where $\tilde{C}_m$ is the concentration at the melting surface and $[\tilde{T}_{\tilde{x}}]$ denotes the jump in $\tilde{T}_{\tilde{x}}$ across this surface. The boundary conditions for the system are given by

$$\tilde{C} \to \tilde{C}_u, \; \tilde{T} \to \tilde{T}_u, \text{ as } \tilde{x} \to -\infty$$

$$\tilde{C} \to 0, \; \tilde{T} \to \tilde{T}_b, \text{ as } \tilde{x} \to +\infty,$$

where the subscripts u and b refer to unburned and burned, respectively. We observe that the burned temperature $\tilde{T}_b$ is derivable from the time-independent solution of the problem as $\tilde{T}_b = \tilde{T}_u + \tilde{\beta}\tilde{C}_u$.

We nondimensionalize by introducing

$$C = \frac{\tilde{C}}{\tilde{C}_u}, \quad = \frac{\tilde{T} - \tilde{T}_u}{\tilde{T}_b - \tilde{T}_u}, \quad t = \frac{\tilde{t}\tilde{U}^2}{\tilde{\lambda}}, \quad x = \frac{\tilde{x}\tilde{U}}{\tilde{\lambda}},$$

$$\phi = \frac{\tilde{\phi}\tilde{U}}{\tilde{\lambda}}, \quad \sigma = \frac{\tilde{T}_u}{\tilde{T}_b}, \quad \gamma = \frac{\tilde{\gamma}}{\tilde{\beta}}, \quad N = \frac{\tilde{E}}{R\tilde{T}_b}.$$

The reference velocity $\tilde{U}$ is the velocity of the uniformly propagating front in the asymptotic limit $N >> 1$. We also introduce the moving coordinate system

$$z = x - \phi(t), \tag{2.3}$$

so that the position of the melting front is fixed at $z = 0$.

In terms of the nondimensionalized quantities, the system (2.1) becomes

$$\begin{aligned} \Theta_t &= \phi_t\Theta_z + \Theta_{zz} + \left(\frac{1}{\alpha(1+\gamma)}\right) gAC\exp\left(\frac{N(1-\sigma)(\Theta - 1)}{\sigma + (1-\sigma)\Theta}\right) \\ C_t &= \phi_t C_z - \left(\frac{1}{\alpha}\right) gAC\exp\left(\frac{N(1-\sigma)(\Theta - 1)}{\sigma + (1-\sigma)\Theta}\right), \end{aligned} \tag{2.4}$$

subject to the boundary conditions

$$\begin{aligned} &C \to 1, \ \Theta \to 0 \text{ as } z \to -\infty \\ &C \to 0, \ \Theta \to 1 \text{ as } z \to +\infty. \end{aligned} \tag{2.5}$$

Note that the boundary condition $C \to 0$ as $z \to +\infty$ follows from (2.4). At the melting surface $z = 0$, the temperature Θ is fixed at Θ_m, and the velocity of the surface is obtained from

$$[\Theta_z] + \gamma C(0)\phi_t = 0. \tag{2.6}$$

The quantity $A = \tilde{\lambda}\tilde{A}/\tilde{U}^2\exp(-\tilde{E}/R\tilde{T}_b)$ is unknown and depends on the (unknown) velocity $\tilde{U}$. It can be determined by finding the solution corresponding to the steadily propagating front. An asymptotic ($N >> 1$) expression for A was derived in [10].

The solution of our problem will be shown to exhibit bifurcation phenomena as we vary the parameter $\mu = \Delta/2(1 - M)$, where $\Delta = N(1 - \sigma)$ and

$M = [1 - \frac{(1 + \gamma)}{\alpha}]\exp[\Delta(\Theta_m - 1)]$.

In order to have a model which is amenable to numerical computation, it is necessary to reduce the problem defined on an infinite domain to one on a finite domain. We introduce finite boundary points $Z_L < 0$, $Z_R > 0$ and require boundary conditions to impose at these points. Highly nonlinear waves are generated near the melting front ($z = 0$) and it is not obvious how to obtain a boundary condition which is absorbing with respect to these waves. We therefore require that the boundaries be placed sufficiently far away so as not to affect the dynamics of the solution. The numerical results were obtained with $Z_R = Z_L = 12$ and we verified that the solution was insensitive to further increases in these values.

At $Z = Z_L$ we imposed the boundary conditions

$$C(Z_L) = 1, \quad \Theta(Z_L) = 0. \tag{2.7}$$

At $Z = Z_R$ only a boundary condition on Θ is required. We tested two boundary conditions. The first was

$$\Theta(Z_r) = 1. \tag{2.8}$$

The second boundary condition was an absorbing boundary condition using the dispersion relation at the analytically predicted Hopf bifurcation point. The analysis in [10, 13] showed that the temperature in the burned region had the form

$$\Theta = 1 + \varepsilon e^{\ell z} e^{i\omega t} \tag{2.9}$$

where $\ell = \frac{1}{2}[1 - (1 + 4i\omega)]$ and ε is a measure of the deviation from the bifurcation point. At the bifurcation point $\omega_0 = 1.029$ and $\ell = -0.30902 -0.63602i$. Guided by (2.9) we can derive a boundary condition by assuming the functional form

$$\Theta = 1 + \varepsilon e^{\ell_1 z} f(\omega t + \ell_2 z),$$

where f is an arbitrary function and $\ell = \ell_1 + i\ell_2$. Differentiating this expression, we obtain

$$\Theta_t = \frac{\omega}{\ell_2}\,\Theta_z - \frac{\ell_1\omega}{\ell_2}(\Theta - 1). \tag{2.10}$$

For the boundary location $Z_R = 12$ we found that (2.7) and (2.10) gave virtually identical results. Using (2.10) we found that Θ oscillated around unity at $Z = Z_R$ but typically the oscillation was of the order $\pm$ 0.003. The computations presented here were obtained using (2.10).

The numerical procedure is based on an adaptive pseudo-spectral method introduced in [3]. For completeness we briefly describe this method. A more complete description of the pseudo-spectral method can be found in [4].

Consider the model equation

$$u_t = au_{xx} + bu_x, \quad -1 \leq x \leq 1. \tag{2.11}$$

In the pseudo-spectral method the approximate solution u is expanded as a finite sum of Chebyshev polynomials

$$u \sim u_J(x,t) = \sum_{n=0}^{J} a_n(t)T_n(x) \tag{2.12}$$

where

$$T_n = \cos(n \cos^{-1} x)$$

is the nth Chebyshev polynomial. The expansion coefficients a_n are obtained by requiring (2.12) to solve (2.11) exactly at the collocation points

$$x_j = \cos(j\pi/J), \quad j = 0,1,\ldots,J. \tag{2.13}$$

The implementation of the collocation procedure proceeds by observing that

$$\frac{\partial^2 u_J}{\partial x^2} = \sum_{n=0}^{J} a_n(t)T_n''(x) = \sum_{n=0}^{J} b_n(t)T_n(x), \tag{2.14}$$

where the coefficients $\{b_n\}$ are related to the coefficients $\{a_n\}$ by a well-known recursion relation. Further details can be found in [4].

In the computations, the intervals $[Z_L, 0]$ and $[0, Z_R]$ are each mapped onto $[-1, 1]$ and the solutions are updated in time within each interval. The velocity of the melting front ϕ_t is determined from (2.6). To determine Θ, C and ϕ_t, a semi-implicit time differencing scheme is used, which is described in detail in [3].

The temperature field exhibits a very rapid variation in a small region behind the melting front. This is the reaction zone in which the reaction term is significant, and outside of which it is not. For the pulsations considered here both the location and width of the reaction zone can vary dynamically.

In order to improve the effectiveness of the pseudo-spectral method behind the melting front, an adaptive procedure was developed in [3]. In this method we introduce a family of coordinate transformations

$$x = q(s,\underset{\sim}{\alpha}) : [-1,1] \to [-1,1], \tag{2.15}$$

where $\underset{\sim}{\alpha}$ is a parameter vector chosen so that in the new coordinate system the weighted second Sobolev norm

$$I(\underset{\sim}{\alpha}) = \int_{-1}^{1} w(s) \; (A \; |u_{ss}|^2 + B \; |u_s|^2 + C \; |u|^2) ds \tag{2.16}$$

is minimized. Here $w(s) = (1-s^2)^{-1/2}$ is the Chebyshev weight function and for our specific computation we have taken $A = B = 1$, $C = 0$. The functional (2.16) is minimized whenever ϕ_t changes by more than a prescribed amount, and the solution interpolated to the new coordinate system where the integration proceeds. For the computations presented here the coordinate transformation

$$q(s,\underset{\sim}{\alpha}) = \frac{4}{\pi} \tan^{-1}\{\alpha \tan [\frac{\pi}{4} (s-1)]\} + 1, \tag{2.17}$$

with $\alpha > 0$, is sufficient, although more general transformations can be used (see [2]). For this problem, (2.16) was used as an indicator of the numerical errors. A different functional, which appears to be more effective, was developed and implemented in [2] for a problem of gaseous combustion. Finally we note that adaptive finite difference methods for combustion problems are presented in [7, 8, 18].

We now describe the behavior of the solutions to the model (2.4)-(2.6), obtained from our numerical computations. The computed pulsating solutions were obtained by solving the time-dependent equations until a steady-state solution was achieved. As a result only stable solutions can be computed. The following parameters were fixed for all of the computations: $N = 50$, $\alpha = 1.7$, $\gamma = 0.5$ and $\Theta_m = 0.8$. The bifurcation parameter μ was changed by varying σ. We have considered values of σ in the range $0.8222 < \sigma < 0.8355$. This corresponds to a variation in μ between $4.208 < \mu < 4.535$. The analytically predicted bifurcation, using a δ function reaction term, occurs at $\mu = 4.236$. For the Arrhenius reaction term with the above parameters, sinusoidal oscillations first appear at a value of μ between 4.270 and 4.281.

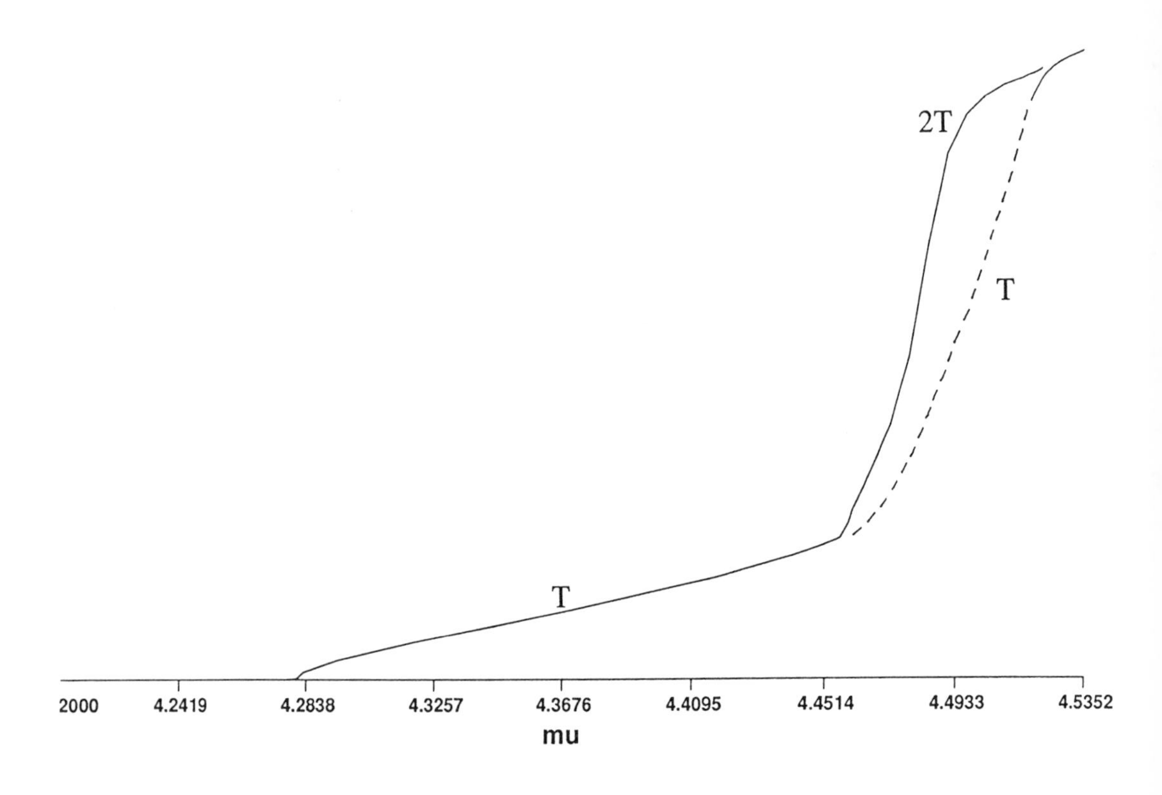

Figure 1(a) Solution branches: Θ_{max} plotted against μ.

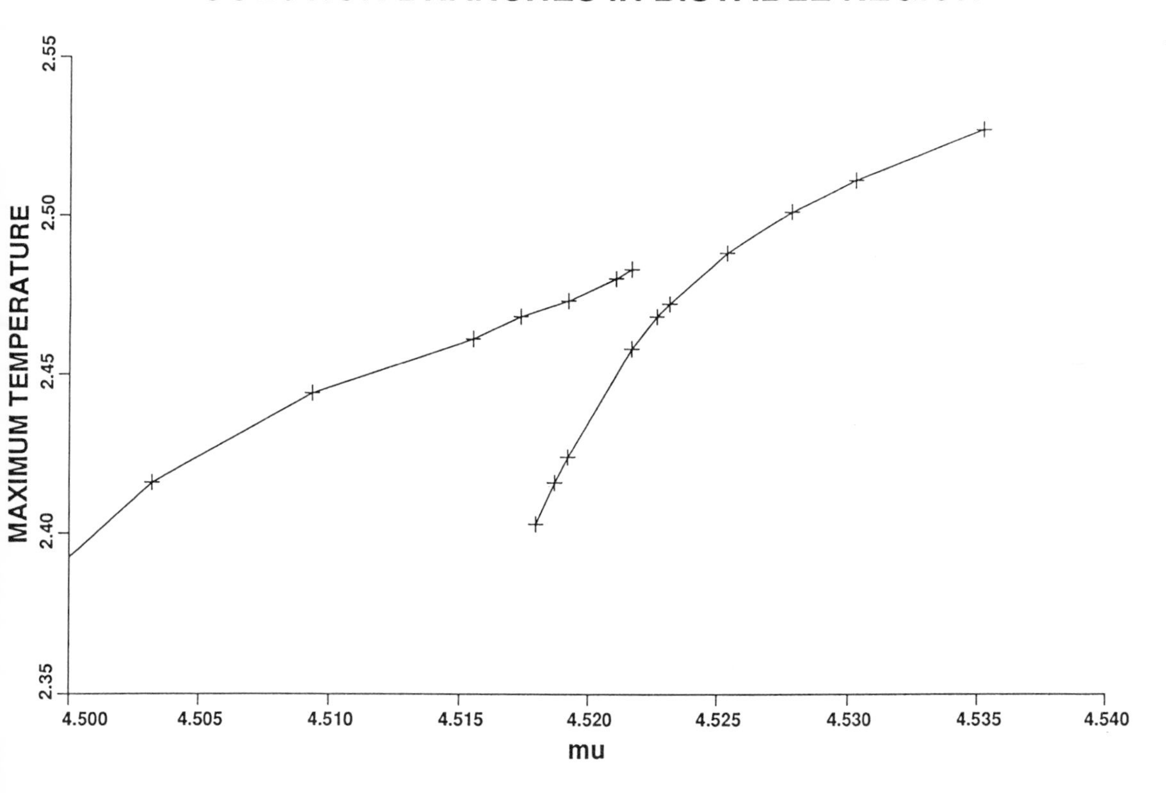

Figure 1(b) The region is expanded to illustrate the bistable behavior

In the bifurcation diagram given in figure 1a, we summarize the different solution branches that have been found. In this bifurcation diagram the maximum value of Θ over one cycle is plotted against μ. Three solution branches are indicated. On the first branch, corresponding to $\mu < \mu_1$, $\Theta_{max} = 1$ and there is no pulsation. This branch corresponds to uniformly propagating reaction waves. On the second branch, corresponding to $\mu_1 < \mu < \mu_2$, the solution is periodic with period $T(\mu)$ and Θ_{max} increases as indicated. On the third branch corresponding to $\mu_2 < \mu < \mu_3$, the solution has become doubly periodic with period 2T. For $\mu > \mu_3$ solutions are again T periodic. These solutions can be continued to values of $\mu < \mu_3$. Thus there is a value μ^* with $\mu_2 < \mu^* < \mu_3$ such that $\mu^* < \mu < \mu_3$ is an interval of bistability in which the T periodic and 2T periodic solutions stably coexist, each with its own domain of attraction. We conjecture that the T periodic branch which exists for $\mu^* < \mu$ is a continuation of the T periodic branch which exists for $\mu_1 < \mu < \mu_2$ and that the portion of this branch (represented by the broken curve in figure 1a) which exists for $\mu_2 < \mu < \mu^*$ corresponds to unstable T periodic solutions. This is a conjecture since our numerical method can only compute stable branches of solutions.

A detailed illustration of the various solution types is presented below. We first discuss the nature of the transitions between the different branches. In general the solution near transition points is difficult to compute, since the equilibration time, i.e. the time for transients to decay, becomes very long. In addition numerical computations may not exhibit sharp bifurcation points even though such points may be present in the underlying analytical model. This is due to perturbations inherent in the numerical discretization which lead to the effect of imperfect bifurcation [12]. Therefore we did not attempt to determine the exact numerical values of the transition points. Our results are based on solutions which were validated by increasing the number of collocation points, i.e. increasing the resolution of the calculation.

Our computational results lead to the following conclusions about the nature of the transitions. The transition between the first two branches corresponds to a Hopf bifurcation at a value μ_1 with $4.270 < \mu_1 < 4.281$. Such a bifurcation was suggested by the linear stability analysis in [10] with $\mu = 4.236$ for a model with a δ function reaction term. For μ close to μ_1, with $\mu > \mu_1$, the solution is nearly sinusoidal. As μ increases the pulsations develop sharp and narrow spikes so that they take on the character of relaxation oscillations.

The transition between the second and third branches appears to occur via a supercritical period-doubling secondary bifurcation at $\mu = \mu_2$ with $4.456 < \mu_2 < 4.459$. We conclude this because the solutions appear to approach the singly periodic solutions continuously as $\mu \to \mu_2$ from above. In addition the equilibration times become very long as $\mu \to \mu_2$.

We now describe the return to the T periodic branch. Stable period-doubled solutions are not found for $\mu > \mu_3$, with $4.521 < \mu_3 < 4.523$. Stable T periodic solutions exist for $\mu > \mu^*$ with $4.515 < \mu^* < 4.518$ and they persist for $\mu > \mu_3$. As stated above, we conjecture that these solutions are part of the same solution branch as the original T periodic solution for $\mu_1 < \mu < \mu_2$, and that the portion of this branch given by $\mu_2 < \mu < \mu^*$ corresponds to unstable solutions. We observe that there is a region of bistability for $\mu^* < \mu < \mu_3$ where both singly and doubly periodic solutions are stable, each having its own domain of attraction. A blow-up of this region is shown in figure 1b. The T periodic branch has been followed up to $\mu = 4.53516$. No additional period doublings or other transitions have been found in this region, though they may occur for other parameter values.

We now give a detailed description of the solutions that have been computed. We found sinusoidal oscillatory solutions near $\mu = \mu_1$, which developed into relaxation oscillations as μ increased. These relaxation oscillations were characterized by a slow movement of the reaction zone followed by a very rapid movement during which the temperature spiked over a very short time interval. As μ is increased further the spikes occur over progressively shorter intervals in time during which the temperature at the spike increases dramatically.

We illustrate this behavior in figures 2a-2g where we plot Θ as a function of t at a fixed value $z(\mu)$. For each vlaue of μ, $z(\mu)$ is chosen to be close to the point where the maximum temperature in both space and time occurs. We note that, by evaluating the global Chebyshev expansion at the given point, we can compute the solution at any given value of z, even though the collocation points adaptively change in time. We observe that the reaction zone moves closer to the melting surface ($z = 0$) as μ increases.

In figure 2a we consider the case $\mu = 4.2597$. In this case there is no pulsation and the solution is exhibited at the arbitrarily chosen point $z = 0.5$ for 20 time units. In figures 2b and 2c we consider $\mu = 4.294$ ($z = 0.5$) and $\mu = 4.454$ ($z = 0.2$). These are singly periodic solutions. The growth of the temperature spike is apparent as is the narrowing of its duration.

In figures 2d and 2e we consider $\mu = 4.466$ ($z = 0.123$) and $\mu = 4.51918$ ($z = 0.026$). The figures illustrate the extremely rapid growth that occurs along the doubly periodic solution branch. At $\mu = 4.51918$ there is bistability and a stable singly periodic solution also exists. This solution is shown in figure 2f ($z = 0.026$). Finally in figure 2g we illustrate the case $\mu = 4.5352$ ($z = 0.026$) along the second stable part of the singly periodic solution branch.

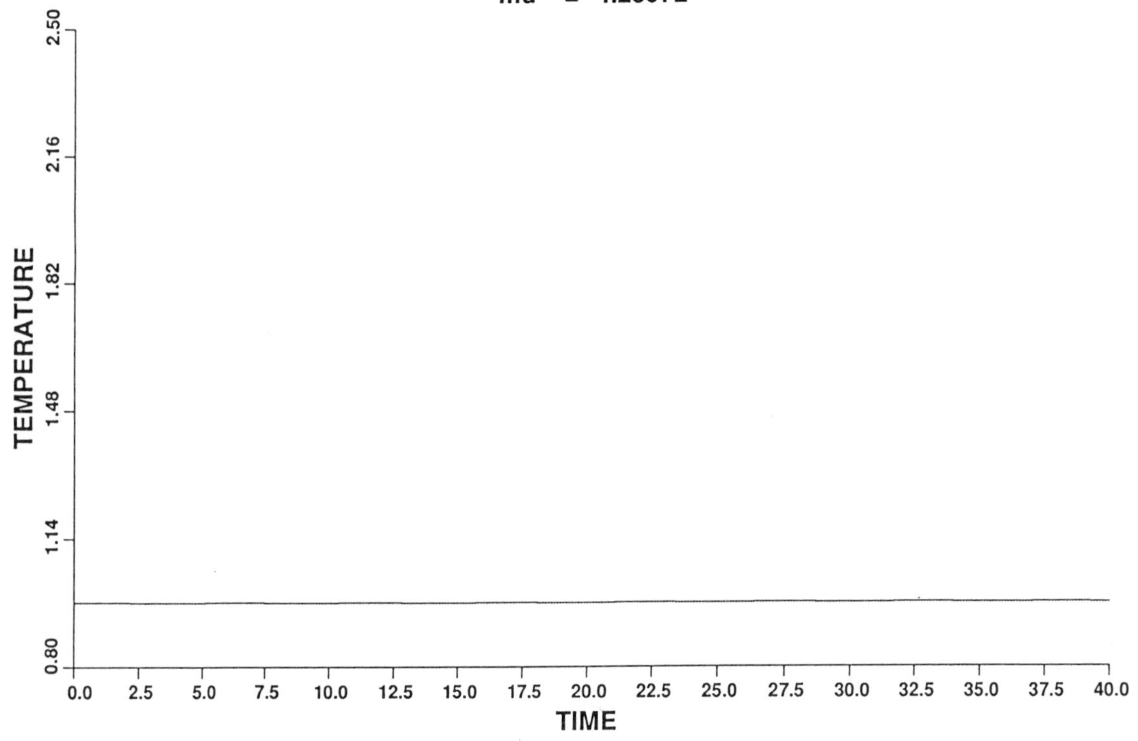

Figure 2(a) Θ at $z = 0.25$, $\mu = 4.260$, steady branch.

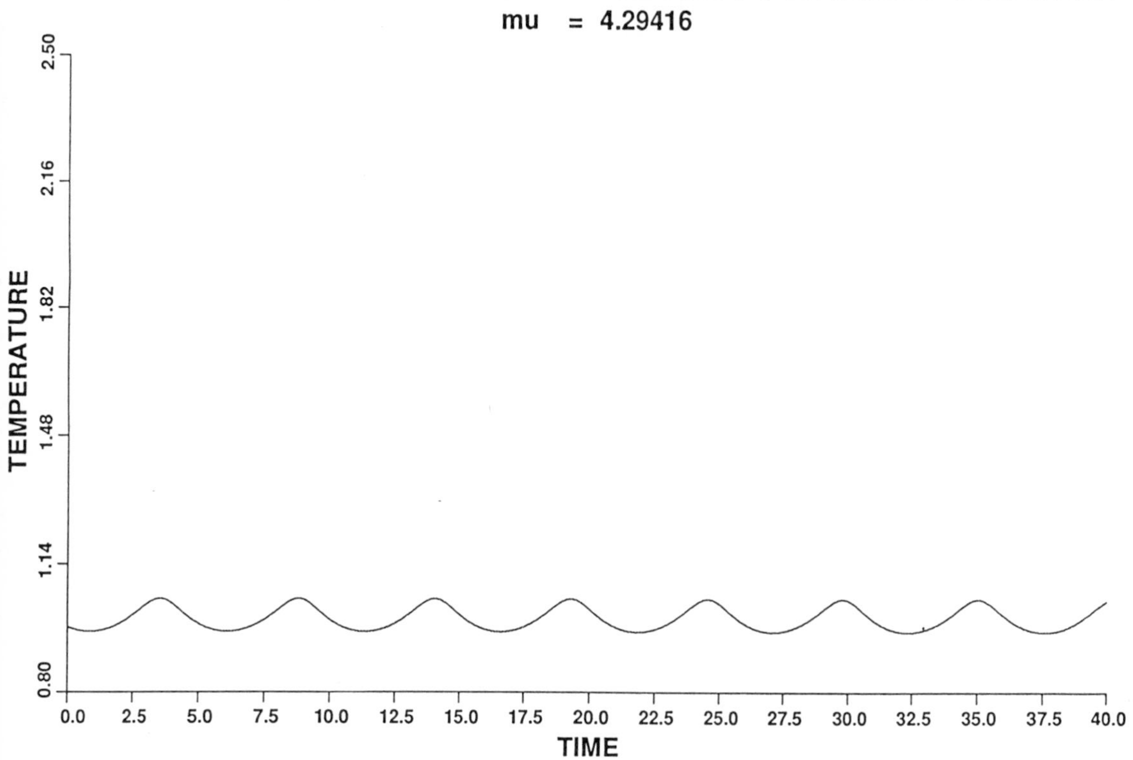

Figure 2(b) Θ at $z = 0.5$, $\mu = 4.294$, singly periodic solution branch.

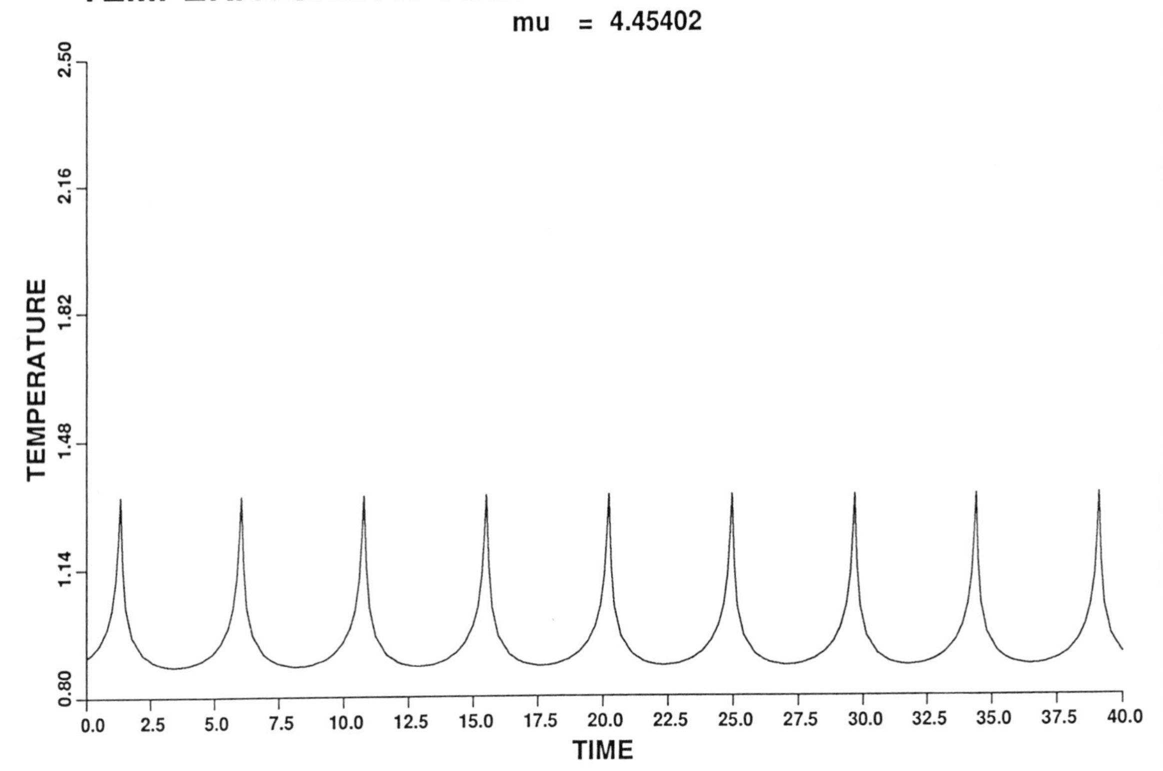

Figure 2(c) Θ at $z = 0.20$, $\mu = 4.454$, singly periodic solution branch.

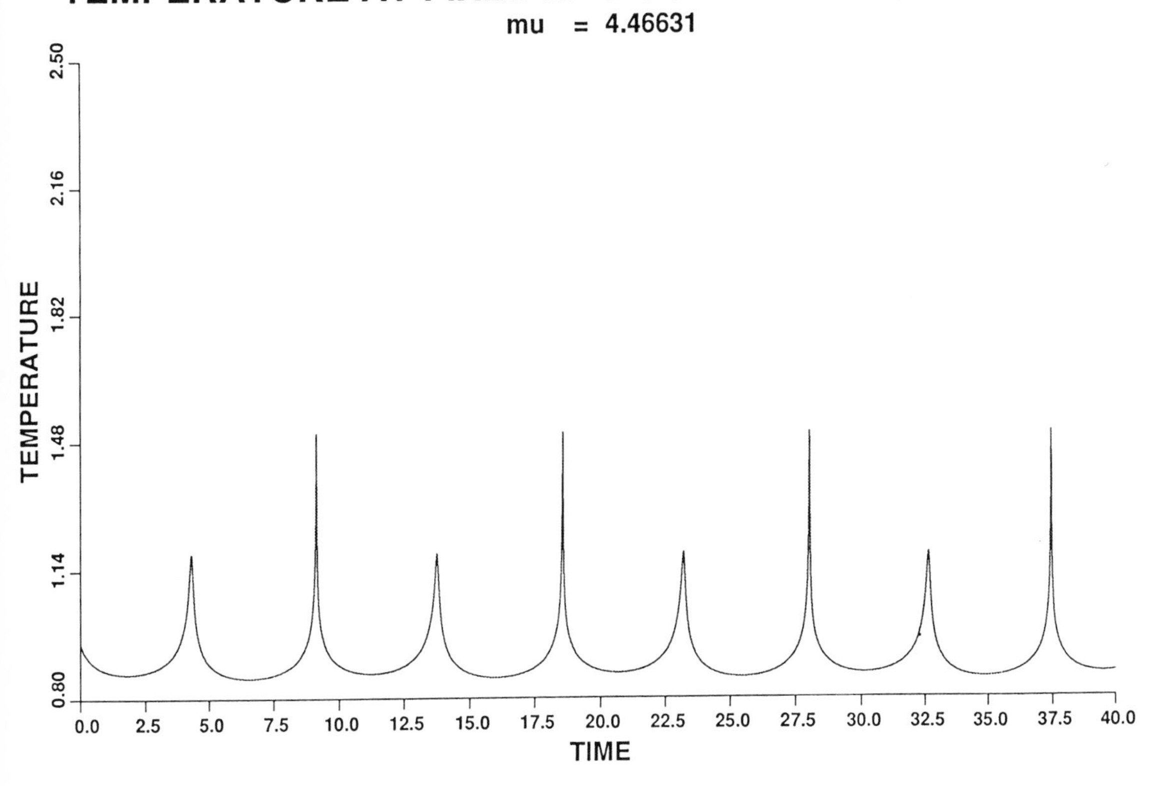

Figure 2(d) Θ at $z = 0.12$, $\mu = 4.466$, doubly periodic solution branch.

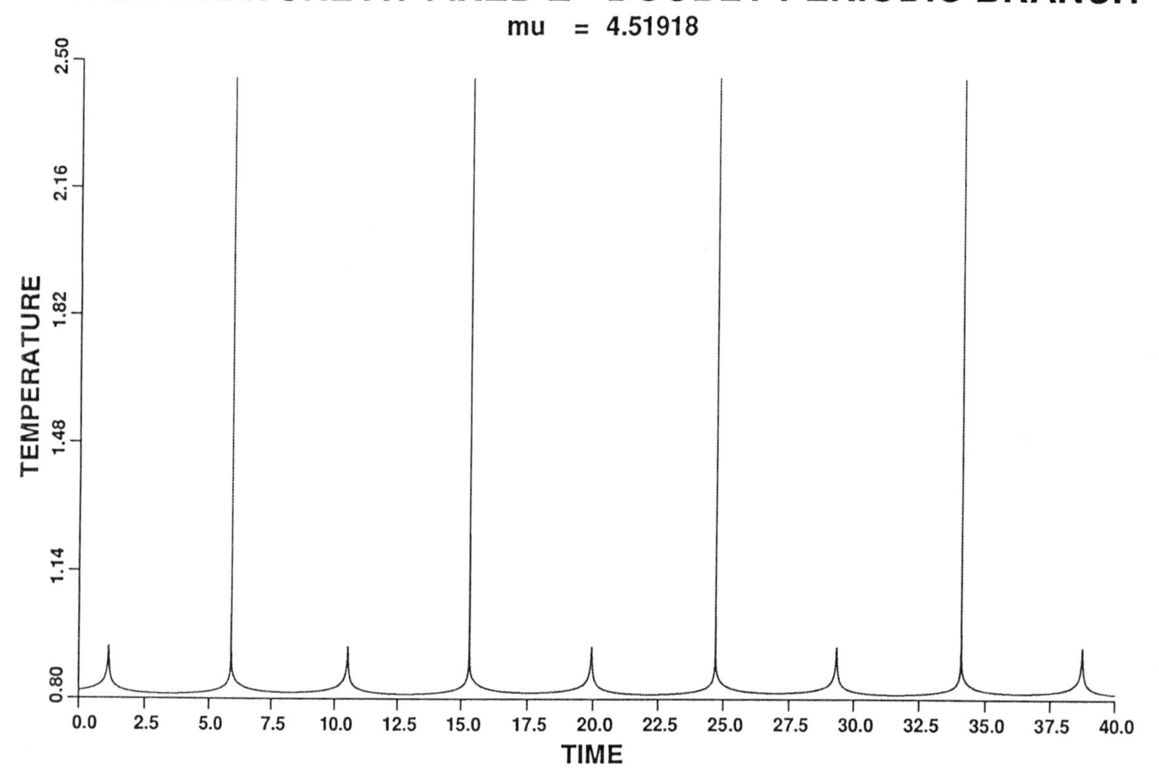

Figure 2(e) Θ at $z = 0.026$, $\mu = 4.51918$, doubly periodic solution branch.

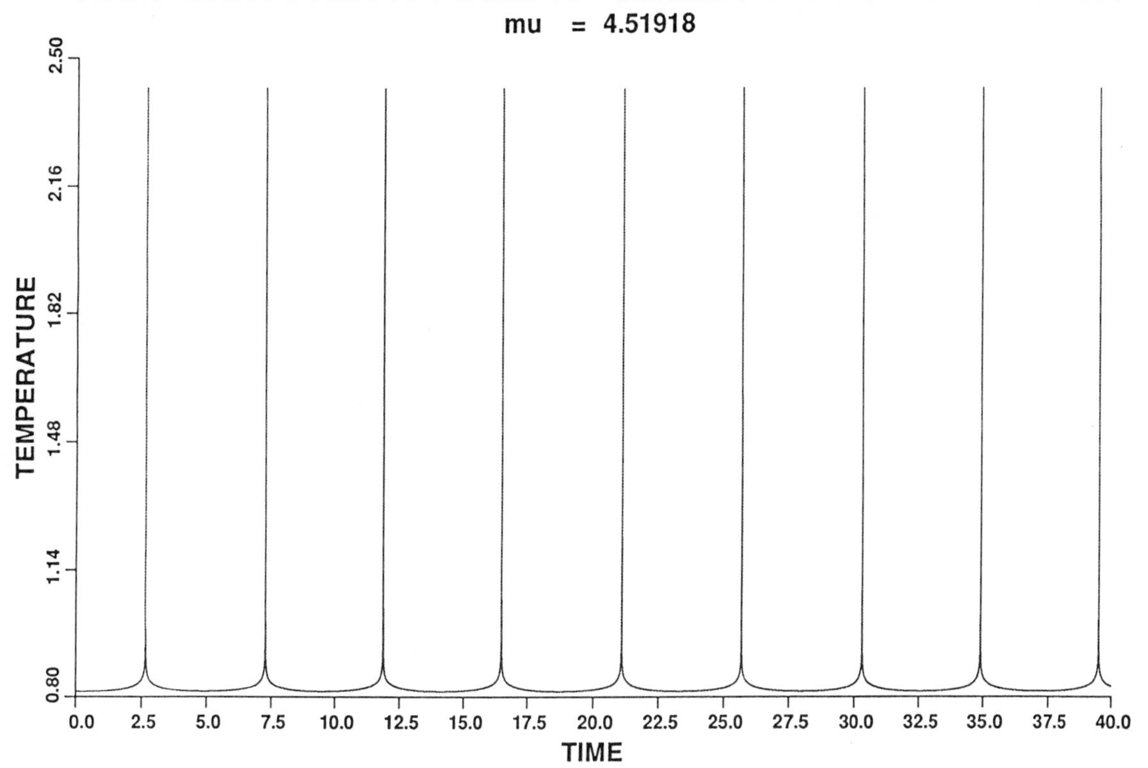

Figure 2(f) Θ at $z = 0.026$, $\mu = 4.51918$, singly periodic solution branch.

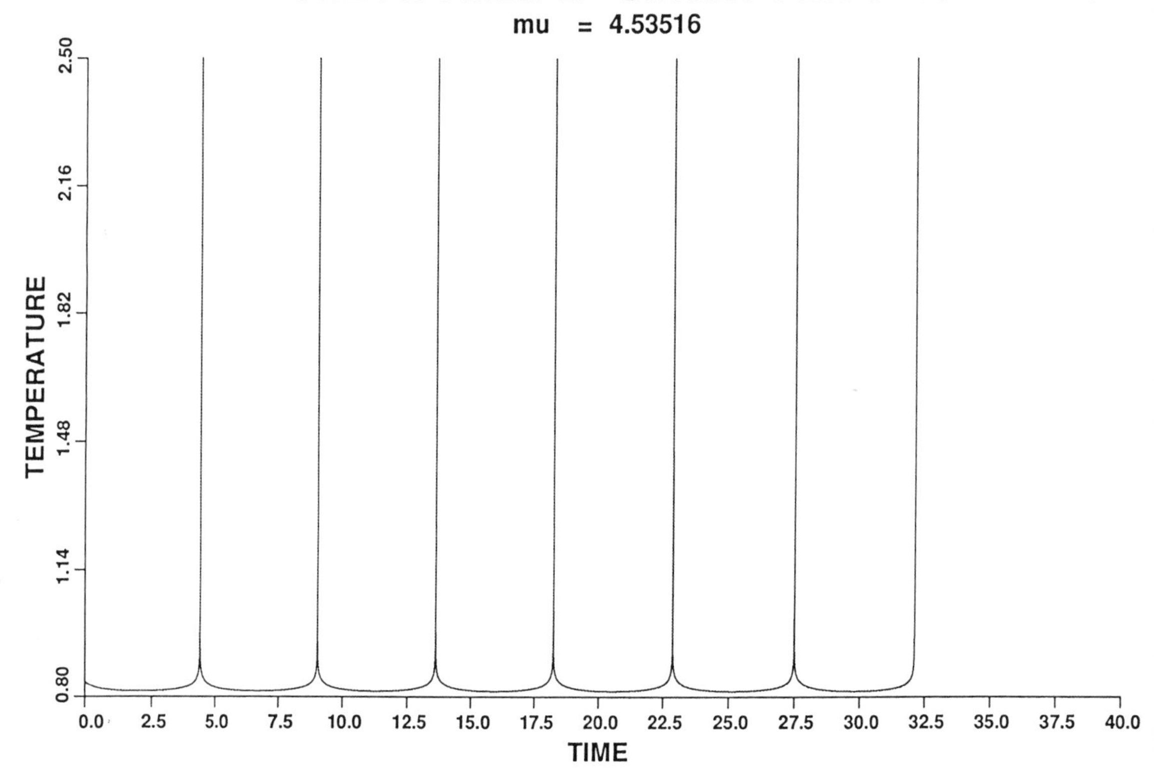

Figure 2(g) Θ at $z = 0.026$, $\mu = 4.5352$, singly periodic solution.

The figures illustrate the rapid growth of the temperature spike as μ increases, and its extremely short duration. In figure 2g the temperature rises from $\Theta = 1.97$ to $\Theta = 2.51$ over a time interval of 1.6×10^{-5}. Very high numerical accuracy was required to resolve these spikes and to verify the periodicity of the pulsation.

The velocity $\dot{\phi} = \phi_t$ of the melting front undergoes a pulsation similar to that of the temperature. We illustrate this for $\mu = 4.51918$ for the doubly periodic solution (figure 3a) and the singly periodic solution (figure 3b). It can be seen that $\dot{\phi}$ increases by more than two orders of magnitude over a cycle. The temperature spike occurs slightly before the spike in $\dot{\phi}$. After the temperature spike, there is a rapid diffusion of heat into the colder material. This surge of heat then results in a rapid movement of the location of melting, as illustrated in figures 3a and 3b.

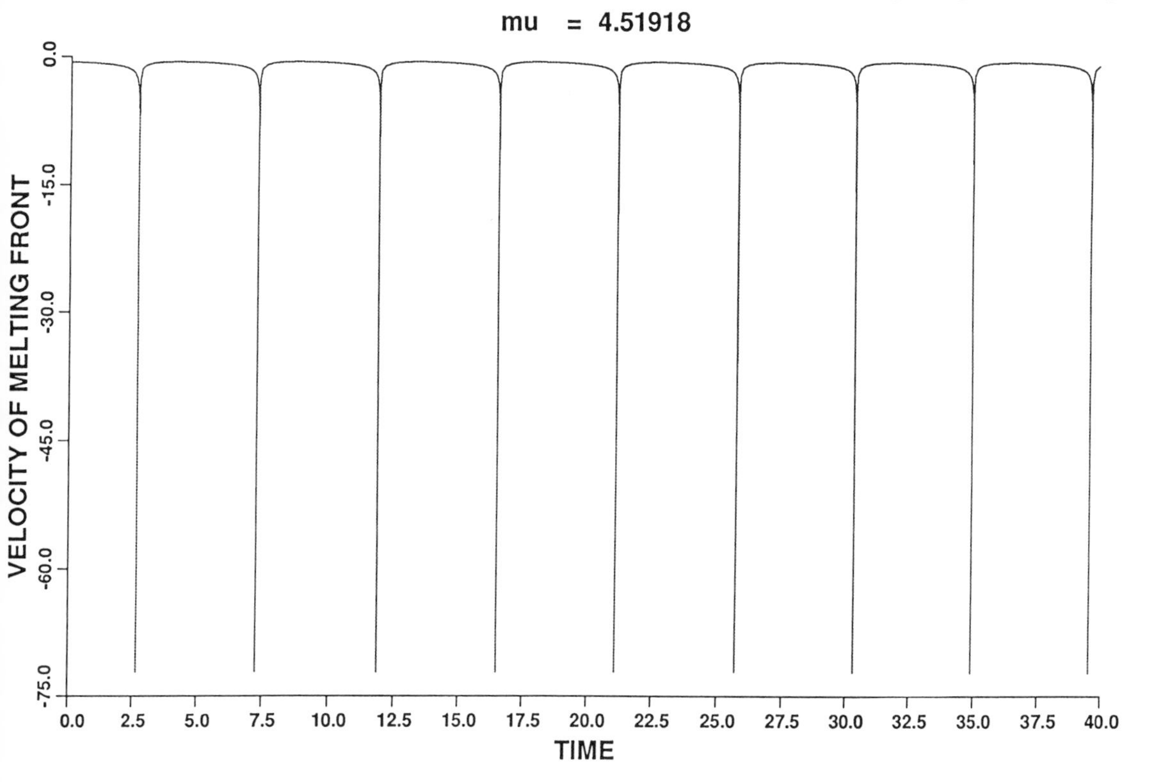

Figure 3(a) $\dot{\phi}$, $\mu = 4.51918$, doubly periodic solution branch.

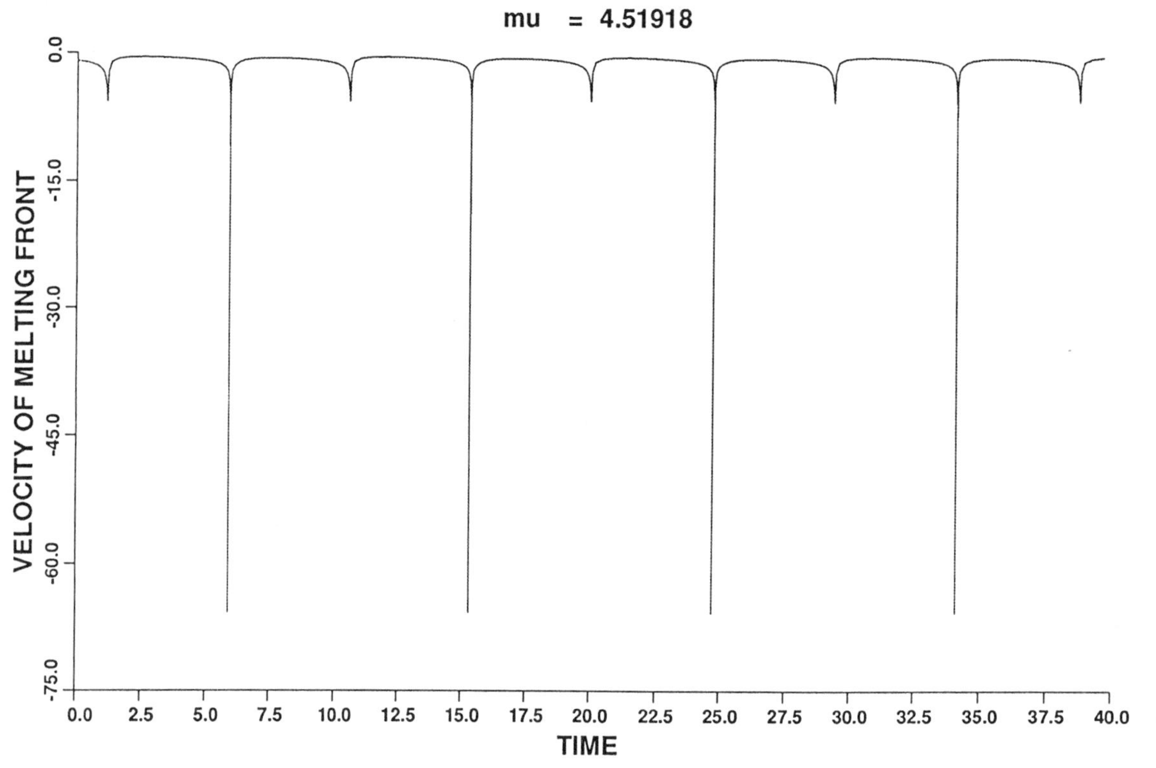

Figure 3(b) $\dot{\phi}$, μ = 4.51918, singly periodic solution branch.

In figures 4a-4d we plot spatial profiles of the temperature. In figure 4a, where there is no pulsation, the solution is shown at the final time of the computation. In figures 4b-4d the solution is shown at the time that $\dot{\phi}$ spikes. The figures illustrate the localized nature of the reaction zone and the extremely rapid spatial variation of the solution. There is a sharpening of the temperature profiles, and a rapid growth of the maximum temperature, as μ increases along the bifurcation branches. We observe that the basic structure of the solution persists for all the values of μ considered. During the slow part of the pulsation, the temperature varies much more gradually.

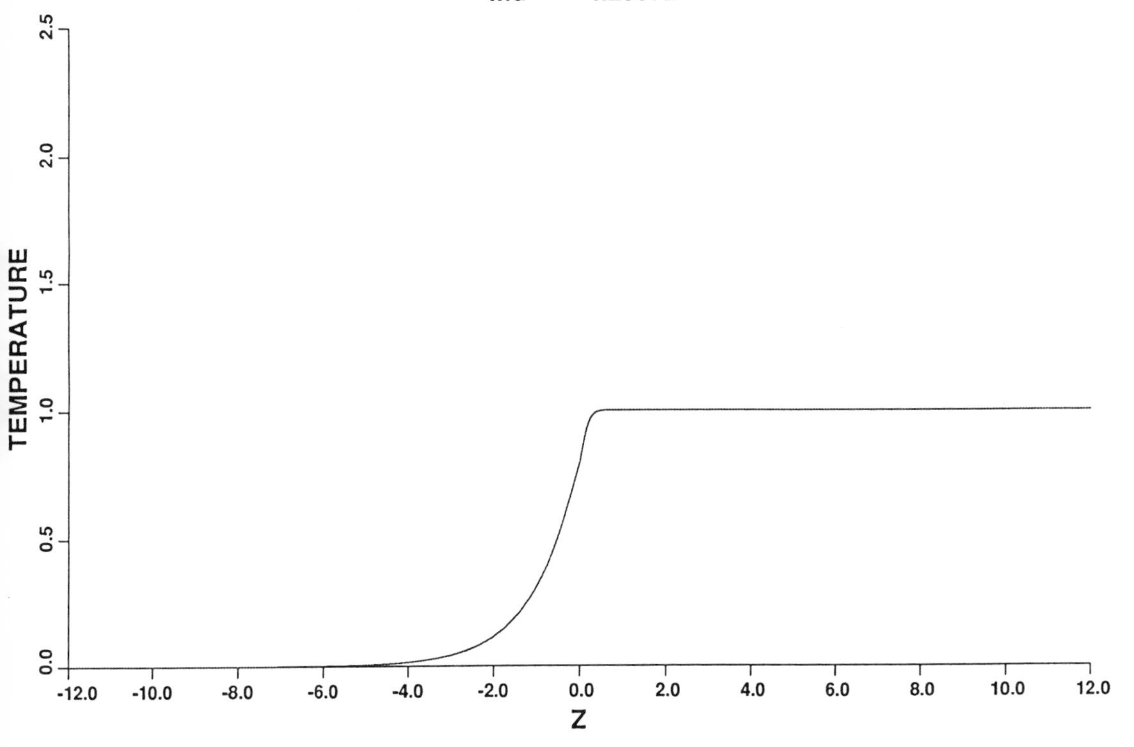

Figure 4(a) Spatial profile of Θ, μ = 4.260, no pulsation.

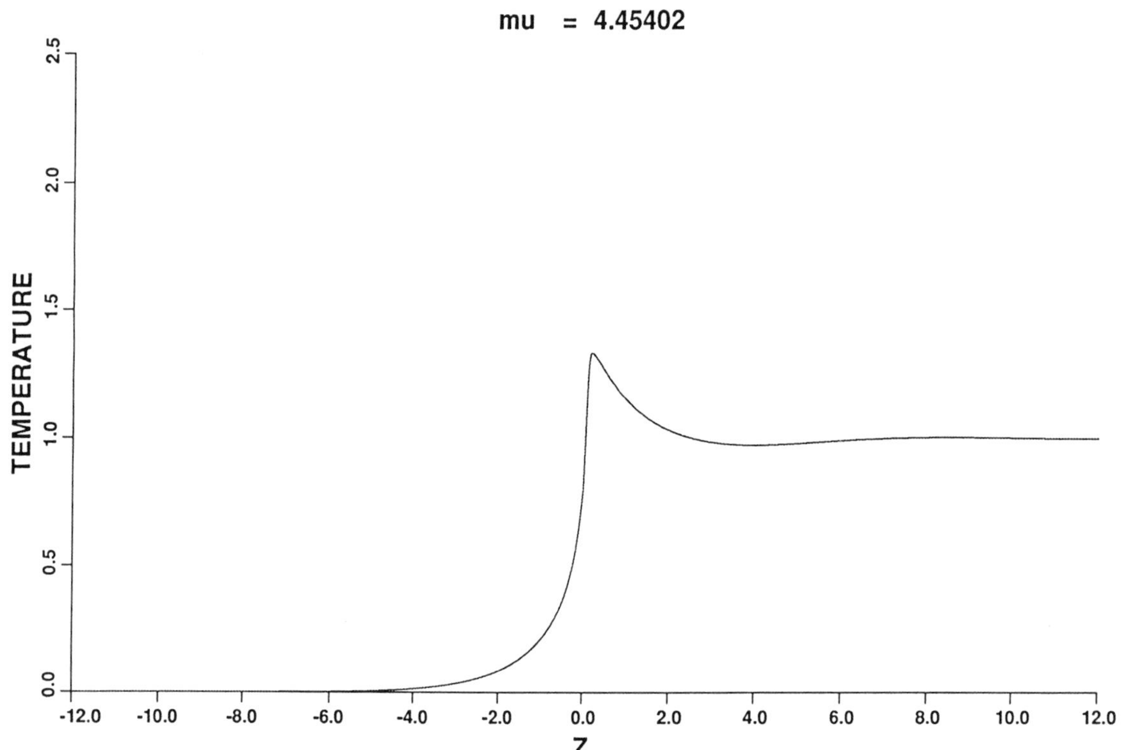

Figure 4(b) Spatial profile of Θ when $|\dot{\phi}|$ is maximum, $\mu = 4.454$, singly periodic solution branch.

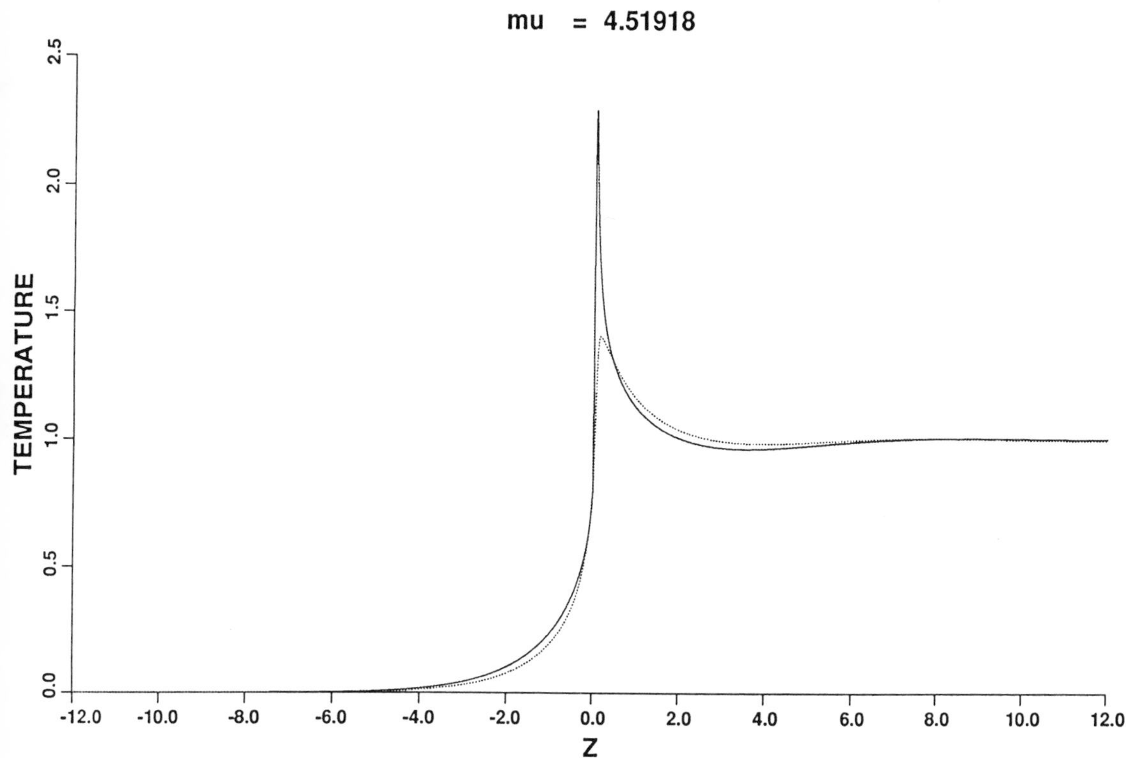

Figure 4(c) Spatial profile of Θ when $|\dot{\phi}|$ is maximum, μ = 4.51918, doubly periodic solution branch.

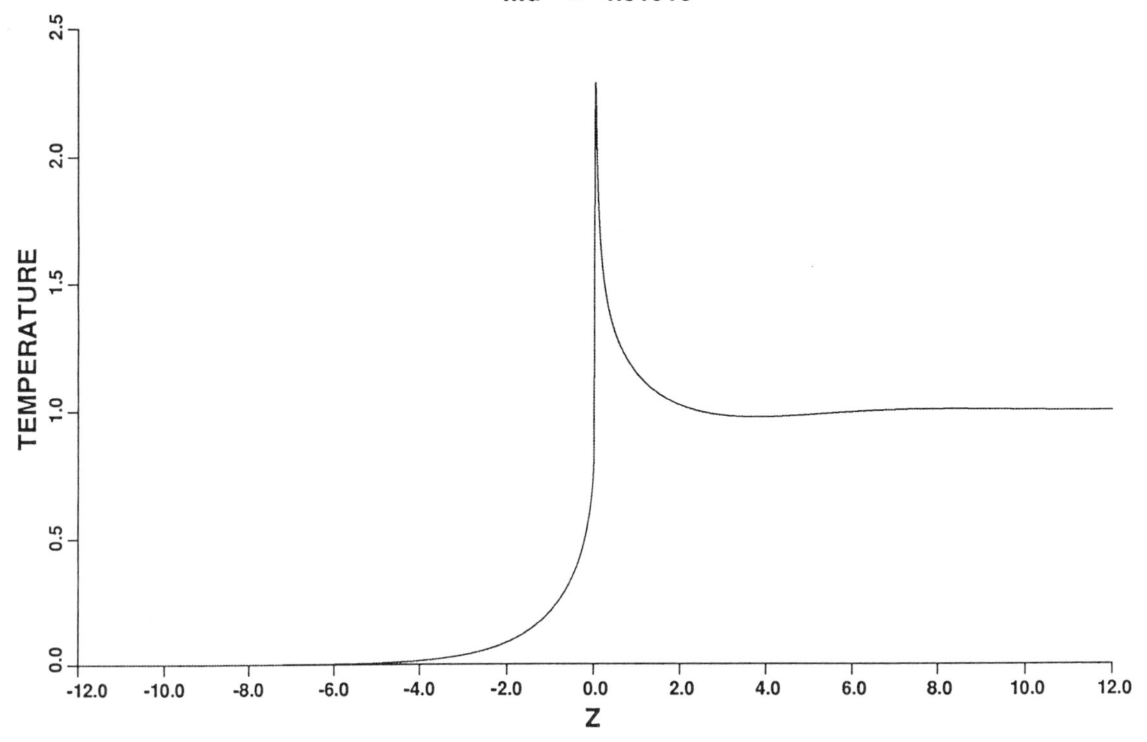

Figure 4(d) Spatial profile of Θ when $|\dot{\phi}|$ is maximum, $\mu = 4.51918$, singly periodic solution branch.

Figure 5 illustrates temperature profiles for a singly periodic solution at four different times. In the figure, t_3 is close to the time of the temperature spike, t_1 corresponds to the minimum of $|\dot{\phi}|$, while t_2 and t_4 are points at which $|\dot{\phi}| = 1$. A corresponding graph for a doubly periodic solution is similar.

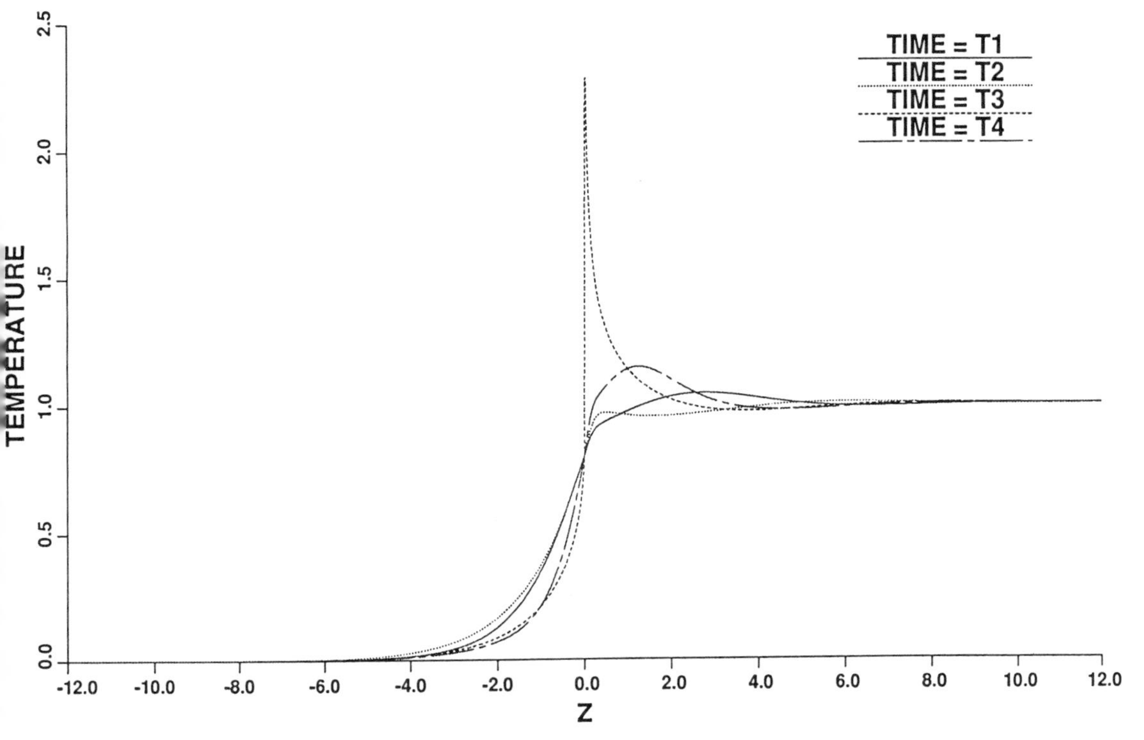

Figure 5 Spatial profiles of Θ at four different times, $\mu = 4.51918$, singly periodic solution branch.

3. GASEOUS FUEL COMBUSTION

We now consider the behavior of cellular flames stablized by a line source of fuel [20]. Experimental observation of laminar flames show that in certain gaseous mixtures smooth flames often break up into cells [24]. Cellular flames, sometimes referred to as wrinkled flames, are characterized by periodic pointed crests along the flame front pointing in the direction of the combustion products. The pointed crests are connected by smooth troughs that are convex toward the fresh fuel. The temperature is higher at the troughs which therefore appear brighter and lower at the crests which appear darker. It is believed that the development of cellular flames is a stage in the transition from laminar to turbulent flame propagation.

In typical combustion problems the activation energies are large. As a result the reaction terms in the governing equations are significant only in a small region, termed the reaction zone. In this region the fuel is consumed and the products of combustion are formed. Ahead of the reaction zone the temperature is too low to sustain the reaction, while behind the reaction zone the fuel is essentially depleted and no reaction can occur. The extent of the reaction zone is $O(1/N)$ where N is an appropriately non-dimensionalized activation energy. In the limit of infinite activation energy the reaction zone shrinks to a surface called the flame front. Across the flame front temperature and concentration are continuous but have discontinuous normal derivatives [26]. Thus cellular flames appear as a wrinkling of the flame front.

We consider problems where the reaction is governed by a deficient component which is consumed in the reaction zone. The specific problem studied here is that of a flame stablized by a line source of fuel of strength $2\pi\kappa$. This problem was analyzed in [25] where a time-independent, axisymmetric solution valid in the limit of infinite activation energy was found. This solution is referred to as the basic solution. The analysis revealed that stationary cellular flames arose as bifurcations from the basic solution, when the Lewis number L, the ratio of heat conductivity to diffusivity of the limiting component of the reaction, was less than a critical value $L_c < 1$. In that case there was a value $\kappa_c(L)$ such that for $\kappa > \kappa_c(L)$ stable, stationary cellular flames existed.

Solutions on the cellular solution branches as κ is increased further into the more fully nonlinear regime are studied by numerical computations. The results of these computations, for a finite value of the activation energy, are presented here. Numerical studies of this problem appear in [20,21,22]. In [21] the transition from an axisymmetric solution to a cellular solution was illustrated as a bifurcation phenomenon as L was decreased with κ fixed. The nonlinear growth of these cells, as L was further decreased, was also illustrated.

In [22] it was shown that for fixed L, as κ increased, transitions could occur between cellular flames of different mode number. A bistable region was found where stable cellular flames of different mode number coexisted at the same parameter values, each with its own domain of attraction. A more complete numerical study of these transitions is presented in [20].

Our results can be best illustrated by referring to figures 6 and 7. In these figures we plot the maximum difference between the temperature and concentration of the cellular solution and the corresponding axisymmetric solution. We find that as κ is increased beyond the first bifurcation point there is a transition to stable, stationary cellular flames of increasing mode number. We have computed stationary cellular solutions with mode numbers increasing from 3 to 6. Around each transition we find an interval of bistability.

Our numerical method is based on integrating the time-dependent equations of the model until a time-independent solution is obtained. As a result, we generally compute stable solutions and cannot give a precise characterization of the mechanism of the transition. In certain circumstances we can compute unstable solutions and illuminate the nature of the transition.

Typically modal transitions occur when cells of one mode number become unstable to angular perturbations of an adjacent mode number. Thus if the original modal solution branch continues beyond the transition point, these solutions can be computed using a time-dependent formulation, provided the computational domain is restricted to exclude the unstable modes. By restricting the angular domain to an angular sector we have been able to compute unstable modal solutions. Based on these computations, we conjecture that the transitions occur via subcritical bifurcations connecting the different unimodal solution curves. In the parameter range considered here, we do not find stable mixed mode solutions, although it is probable that unstable mixed mode solutions exist. For the parameter range considered, we have not found more than two stable cellular solutions coexisting at given parameter values.

To describe the mathematical model, we let a tilde (~) denote a dimensional quantity. We consider the following dimensional quantities: $\tilde{T}_u$, $\tilde{T}_b$, the temperatures of the unburned mixture and the adiabatic flame temperature respectively; $\tilde{C}_u$, the concentration of the deficient component of the unburned reactant (fuel); $\tilde{E}$, the activation energy; $\tilde{R}$, the gas constant; $\tilde{\lambda}$, the thermal diffusivity; and $\tilde{A}$, the rate constant. Nondimensional temperature and concentration are defined by

$$C = \tilde{C}/\tilde{C}_u$$

$$\Theta = (\tilde{T} - \tilde{T}_u)/(\tilde{T}_b - \tilde{T}_u),$$

where $\tilde{T}$ is the temperature and $\tilde{C}$ is the concentration of the deficient component. The nondimensional activation energy is defined by

$$N = \tilde{E}/(\tilde{R}\tilde{T}_b).$$

The spatial and temporal variables are nondimensionalized by

$$t = \frac{\tilde{t}\,\tilde{v}^2}{\tilde{\lambda}}, \quad x_i = \frac{\tilde{x}_i\tilde{v}}{\tilde{\lambda}},$$

where $\tilde{v}$ is the planar, adiabatic flame speed. Assuming an appropriately nondimensionalized flow field $\underset{\sim}{U}$ and one-step irreversible Arrhenius kinetics, the equations of the diffusional thermal model of combustion are [26]

$$\begin{aligned} \Theta_t &= \Delta\Theta - \underset{\sim}{U}\cdot\nabla\Theta + \Lambda C \exp\left[\frac{N(1-\sigma)\ (\Theta-1)}{\sigma + (1-\sigma)\Theta}\right] \\ C_t &= \frac{\Delta C}{L} - \underset{\sim}{U}\cdot\nabla C - \Lambda C \exp\left[\frac{N(1-\sigma)\ (\Theta-1)}{\sigma + (1-\sigma)\Theta}\right]. \end{aligned} \tag{3.1}$$

Here L, the Lewis number, is the ratio of thermal conductivity to the diffusivity of the deficient component. The quantity Λ is called the planar flame speed eigenvalue and depends on the unknown reference velocity $\tilde{v}$. In the limit $M = N(1-\sigma)$ large the asymptotic expansion of Λ is known [23]:

$$\Lambda = M^2 \left[\frac{1}{2L} + O\left(\frac{1}{M}\right)\right]. \tag{3.2}$$

In our computations we use the leading order term in (3.2) to approximate Λ in (3.1). A different value of Λ will alter the spatial and temporal scales but will not change any patterns found for the solutions. The external flow field $\underset{\sim}{U}$ is taken as that of a line source of fuel of strength $2\pi\kappa$, i.e.

$$U = \frac{\kappa}{r}\hat{r}, \tag{3.3}$$

where $\hat{r}$ is a radial unit vector. The boundary conditions are

$$\begin{aligned} &\Theta \to 0(1), \ r \to 0(\infty), \\ &C \to 1(0), \ r \to 0(\infty). \end{aligned} \tag{3.4}$$

It is characteristic of combustion problems that the activation energies are large and the reaction terms are important only in a narrow region called the reaction zone. In the limit $N >> 1$, $(1-\sigma) << 1$, $M >> 1$, the reaction zone shrinks to a surface $r = \Psi(\phi)$, where ϕ is the polar angle, called the flame front, across which the normal derivatives of Θ and C are discontinuous with derived jump conditions [26]. In this limit the axisymmetric, stationary solution

$$\begin{aligned} &\Theta = \left(\frac{r}{\kappa}\right)^{\kappa} + O\left(\frac{1}{M}\right), \quad r \leqq \kappa \\ &\Theta = 1, \qquad r \geqq \kappa \\ &C = 1 - \Theta + O\left(\frac{1}{M}\right) \\ &\Psi = \kappa \end{aligned} \tag{3.5}$$

exists. The effect of the fuel source is to stabilize the front location at $r = \kappa$. The solution (3.5) is called the basic solution.

The stability of (3.5) was analyzed in [25]. The basic solution is stable for L near unity. There exists a critical value of L, $L_c = 1 - O(1/M)$, such that if $L < L_c$ the basic solution is unstable to angular perturbations for κ

sufficiently large, $\kappa > \kappa_c(L)$. A weakly nonlinear analysis showed that small disturbances evolved into stationary cellular flames.

The behavior of the model (3.1) as the parameters move into the more fully nonlinear regime is the subject of this paper. The computations are for finite activation energy. In the numerical computations the boundary conditions (3.4) must be applied at artificial boundary points r_1 and r_2. We apply the boundary conditions

$$\begin{aligned} \Theta(r_2) &= C(r_1) = 1 \\ \Theta(r_1) &= C(r_2) = 0. \end{aligned} \tag{3.6}$$

In the computations presented here we chose $r_1 = 2.4$, $r_2 = 30.4$. The reaction zone was sufficiently far from the boundaries so that no significant sensitivity to the boundary location was observed.

The numerical method is based on an adaptive Chebyshev-Fourier pseudo-spectral discretization. In our work the radial coordinate system is varied adaptively to enhance resolution of the regions of rapid variation [2,3].

The interval $[r_1, r_2]$ is first mapped into the interval $[-1,1]$ by the linear transformation

$$s = \frac{2r}{r_2-r_1} - \frac{(r_1+r_2)}{r_2-r_1} .$$

We then introduce the collocation points

$$\begin{aligned} s_j &= \cos\,(\pi j/J) \quad (j = 0,\ldots,J) \\ \phi_k &= 2\pi k/K \quad (k = 0,\ldots,K-1) \end{aligned} \tag{3.7}$$

and approximate the solution by the Chebyshev-Fourier expansion

$$\Theta = \sum_{j=0}^{J} \sum_{|\ell| \leq K/2} \hat{\Theta}_{j,\ell} T_j(s) e^{i\ell\phi} \tag{3.8}$$

with a similar expansion for C. Here $T_j(s)$ is the jth Chebyshev polynomial

$$T_j(s) = \cos(j \cos^{-1} s).$$

The coefficients $\hat{\Theta}_{j,\ell}$ in (3.8) are obtained by collocation, that is by requiring equations (3.1) to be exactly satisfied at the collocation points. In the pseudo-spectral method $\Theta_{j,k}$ are the unknowns; the expansion (3.8) is used only to compute derivatives at the collocation points.

In order to improve the resolution of the regions of rapid variation near the reaction zone, we adaptively transform the radial coordinate by a coordinate transformation

$$s = q(s;\alpha)$$

where $\underset{\sim}{\alpha}$ is a two-element vector. The specific form of the transformation is

$$q(s;\underset{\sim}{\alpha}) = \frac{4}{\pi} \tan^{-1}[\alpha_1 \tan(\tfrac{\pi}{4}(s'-1))] + 1$$

$$s' = \frac{\alpha_2 - s}{\alpha_2 s - 1}$$

where $\underset{\sim}{\alpha} = (\alpha_1, \alpha_2)$; $\alpha_1 > 0$ and $-1 < \alpha_2 < 1$. The two parameters α_1 and α_2 provide the flexibility to move regions of rapid variation to the boundary (α_2) and then expand these regions (α_1). Other choices of coordinate transformation are possible.

The vector $\underset{\sim}{\alpha}$ (equivalently the radial coordinate system) is chosen adaptively so as to minimize a functional of the solution which measures the spectral interpolation error. This functional was introduced and described in [2]. For a function f which is a weighted combination of Θ and C, we have

$$I(f) = \left\{ \int_0^{2\pi} d\phi \int_{-1}^{1} \frac{ds}{\sqrt{(1-s^2)}} \left[\left(\sqrt{(1-s^2)}\, \frac{\partial}{\partial s} \right)^2 f \right]^2 \right\}^{1/2} . \qquad (3.9)$$

The analysis and computations presented in [2] demonstrate that (3.9) is an effective measure of the spectral interpolation error. Owing to the angular dependence in the reaction zone, it would be more effective to choose $\underset{\sim}{\alpha}$ for each value of ϕ. Such a procedure is currently under development.

The equations are integrated in time until a stationary solution is achieved. The time marching procedure is a semi-implicit scheme using the backward Euler scheme with approximate factorization. To describe the time differencing we consider the model equation

$$U_t = U_{xx} + U_{yy} + r(U).$$

Using a superscript to denote the time levels and the symbols D_x, D_y to denote the approximate second derivative operators in the x and y directions respectively, we have

$$\frac{U^{n+1} - U^n}{\Delta t} = D_x U^{n+1} + D_y U^{n+1} + r(U^n)$$

or if $\delta = U^{n+1} - U^n$,

$$[I - \Delta t D_x - \Delta t D_y]\delta = \Delta t[D_x U^n + D_y U^n + r(U^n)]. \tag{3.10}$$

The matrix on the left-hand side of (3.10) can be approximately factored (up to $O(\Delta t^2)$) as

$$[I - \Delta t D_x - \Delta t D_y] = [I - \Delta t D_x][I - \Delta t D_y] + O(\Delta t^2), \tag{3.11}$$

and the solution is updated in time by inverting the factored matrix on the right-hand side of (3.11). Convergence to a steady state is monitored by examining the maximum of the residual over the grid, i.e. $\max_{j,k}|D_x U^n + D_y U^n + r(U^n)|$. Typically we require the maximum residual to decrease by 6-9 orders of magnitude. Convergence can be very delicate near a modal transition but is fairly rapid away from transition points.

In all of the calculations presented here the following parameters were held fixed:

$$N = 20,\ \sigma = 0.615,\ L = 0.44,\ r_1 = 2.4,\ r_2 = 30.4.$$

Solution branches were computed with as the parameter that was varied.

The first cellular transition was obtained by using as initial data the axisymmetric solution with an angular perturbation. Subsequent solutions were obtained by varying κ using cellular data with a nearby value of κ as initial data.

All of the computed results presented here were obtained with a grid of 141 radial collocation points and 128 angular collocation points. Clearly the effective angular resolution is less for cells of higher mode numbers. These computations were validated by computing solutions on coarser grids and verifying that there was very little change as the grid was coarsened. Generally the most difficult computations occur near transition points where a cellular solution is very weakly unstable to perturbations of an adjacent mode number. Near these points the amplitudes of all modes must be monitored carefully to ensure that there is not a weak instability which could lead to a modal transition if the solution was computed for a sufficiently long time. We have concentrated on localizing the transition points to intervals outside of which we are confident that we have a stable cellular solution. Our procedure is to follow a particular modal solution branch by varying κ. All other modes are monitored to ensure that they are not growing. When we come to a value of κ where another mode shows persistent growth, so that the given modal solution can be judged unstable, we take that value of κ as an upper (or lower) bound on the stable region of the solution branch.

In discussing the solution branches we refer to figures 6 and 7. In these figures we plot the maximum norm difference of the computed temperature (figure 6) and concentration (figure 7) between the cellular and the axisymmetric solution. The computed values are indicated by full circles in the figures. The unstable axisymmetric solution was computed using an axisymmetric version of the computer program so that angular perturbations were not present. The curves in the figure represent five solution branches, the axisymmetric branch and cellular solutions branches of mode numbers 3-6, respectively.

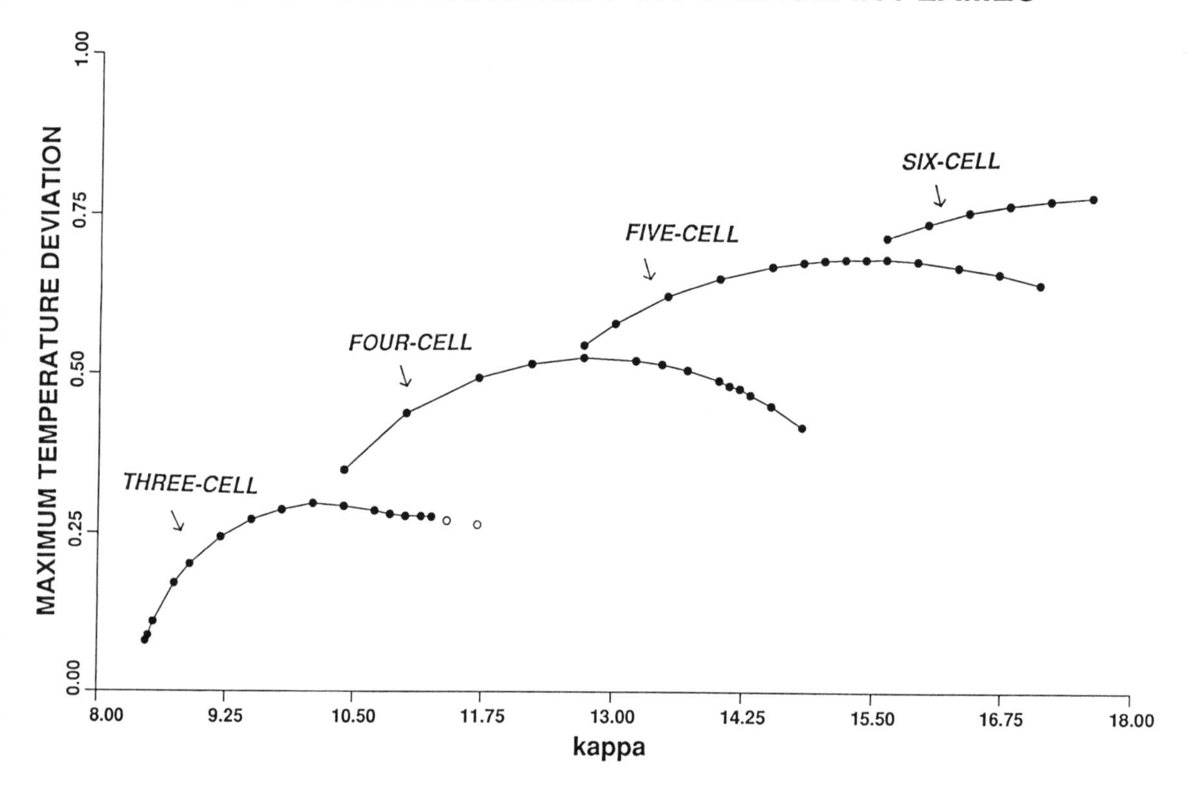

Figure 6 Maximum norm difference of Θ between the computed cellular solution and the axisymmetric solution. Full circles represent actual computed values. Open circles correspond to unstable solutions.

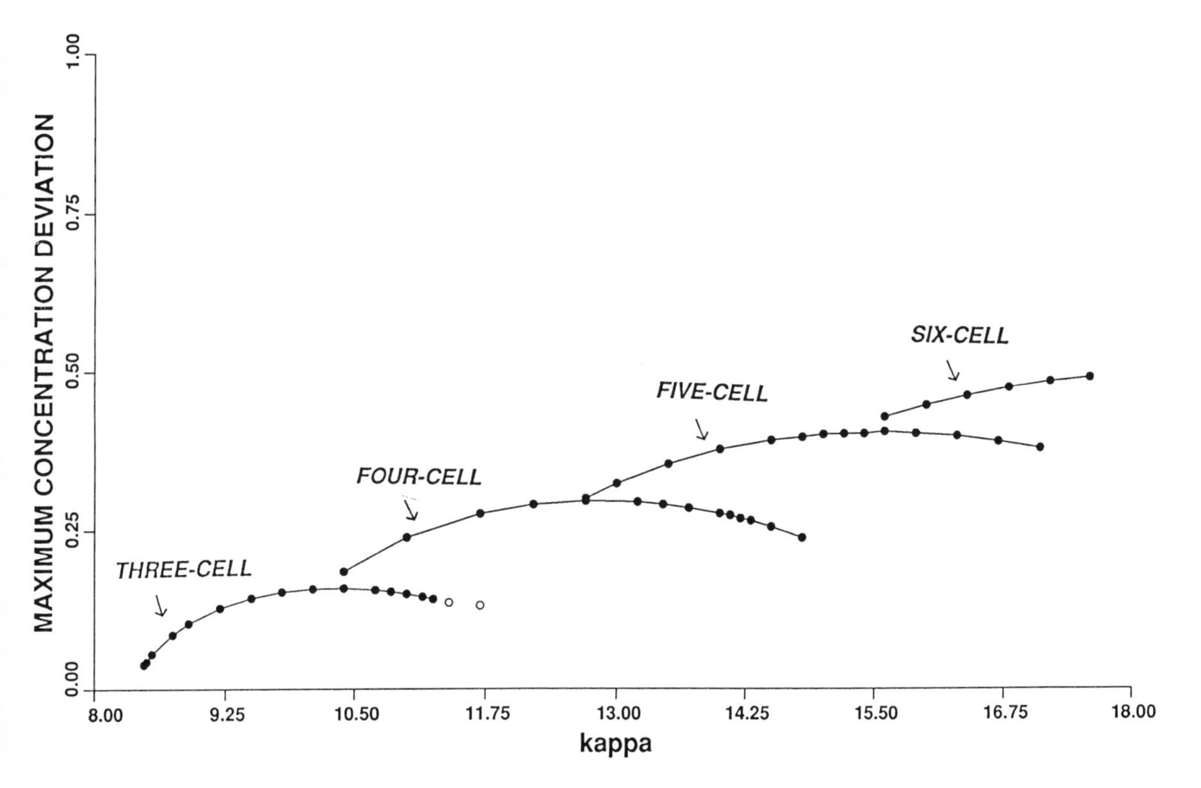

Figure 7 Maximum norm difference of C between the computed cellular solution and the axisymmetric solution. Full circles represent actual computed values. Open circles correspond to unstable solutions.

The axisymmetric solution branch is stable for $\kappa < \kappa_1$ where $8.3 < \kappa_1 < 8.4$. For $\kappa > \kappa_1$ with $\kappa - \kappa_1$ small, the solution with arbitrary initial data evolves to a stationary three-cell. This solution can be computed for $\kappa_1 < \kappa < \kappa_2$ where $11.25 < \kappa_2 < 11.70$. At $\kappa = 11.70$ initial data from a lower value of κ evolved to a stationary four-cell. The four-cell solution branch can be continued for the range $\kappa^1 < \kappa < \kappa_3$ where $10.0 < \kappa^1 < 10.4$ and $14.8 < \kappa_3 < 15.2$. The region $\kappa^1 < \kappa < \kappa_1$ is a region of bistability in which both three- and four-mode cellular flames coexist, each with its own domain of attraction.

For $\kappa > \kappa_2$ initial data from a nearby solution branch evolved to a five-cell solution. The five-cell solution branch can be computed for $\kappa^2 < \kappa < \kappa_3$ where $12.2 < \kappa^2 < 12.7$ and $17.1 < \kappa_3 < 17.5$. A six-cell solution branch exists and has been computed in the range $15.6 < \kappa < 17.6$. An upper limit for this branch has not been determined. It can be seen from the figures that the graphs of these branches are similar. We have not found values of κ where more than two stable solutions coexist.

In figures 8, 9 and 10 we present perspective plots of Θ along the different solution branches. Each figure is for a value of κ in a region of bistability. In figure 8 we plot the three-cell and four-cell solution at $\kappa = 11.0$. In figure 9 we plot the four-cell and five-cell solution at $\kappa = 14.8$. In figure 10 we plot the five-cell and six-cell solution at $\kappa = 15.7$.

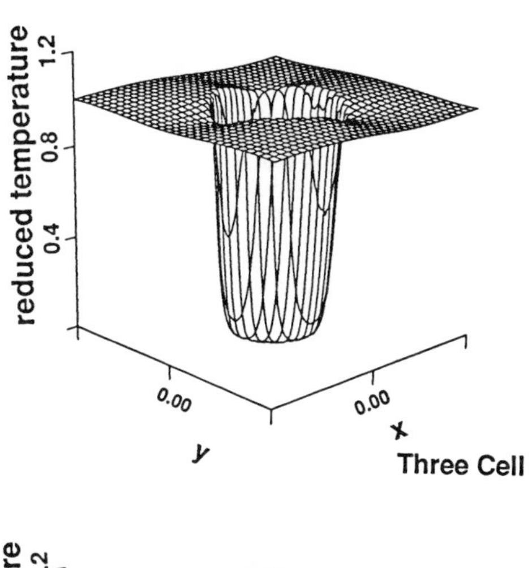

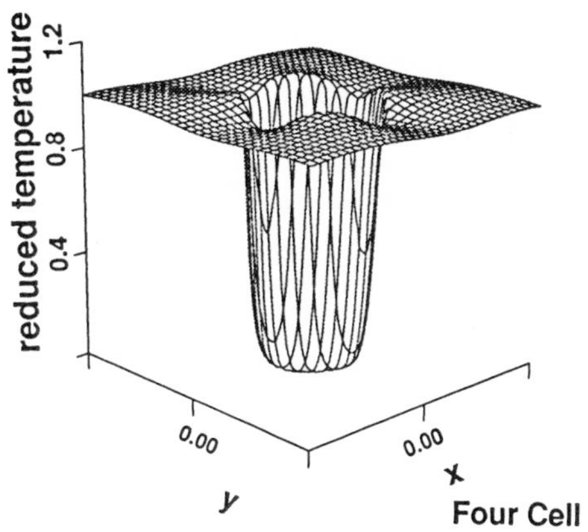

Figure 8 Bistable cellular flames; $\kappa = 11.0$.

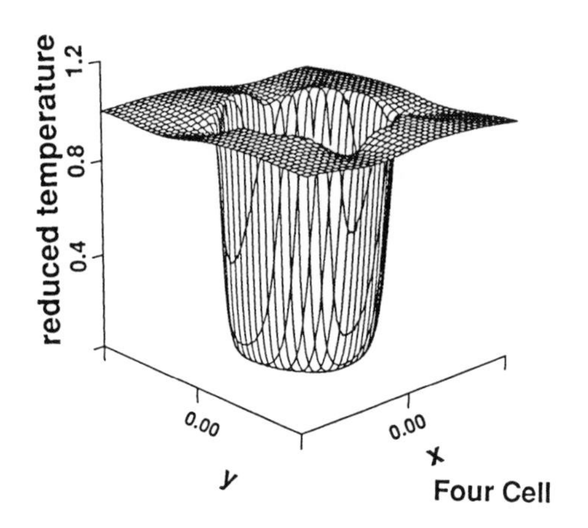

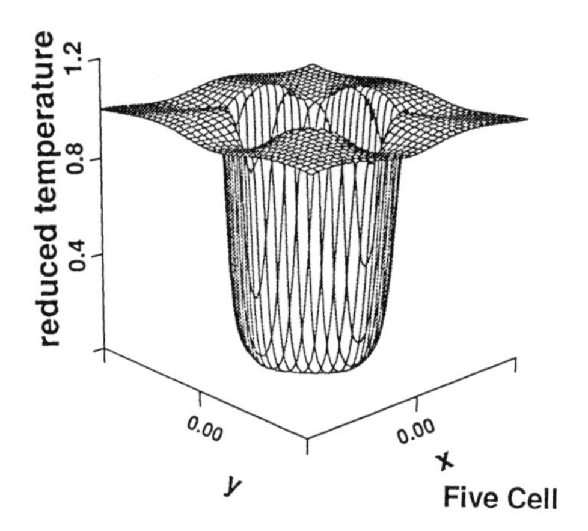

Figure 9 Bistable cellular flames; $\kappa = 14.8$.

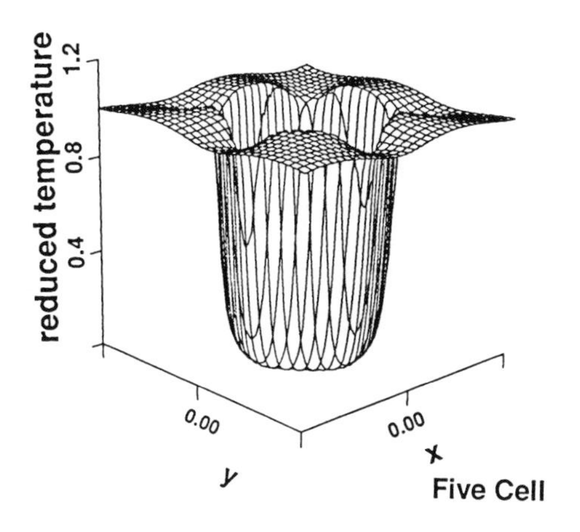

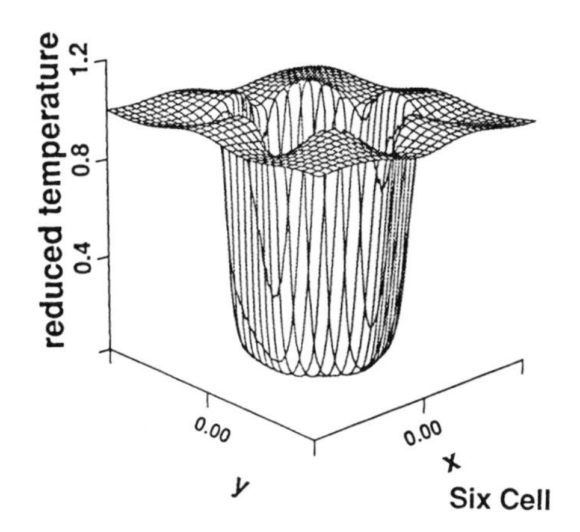

Figure 10 Bistable cellular flames; κ = 15.7.

The figures show a small region over which Θ drops rapidly from unity to zero. In the limit of infinite activation energy this region would shrink to the flame front. The subregion where the temperature begins to drop is called the reaction zone and is the region where the reaction term has the greatest effect on the solution. It is apparent from the figures that the primary effect of the cellular solution is to cause a wrinkling of the reaction zone. The wrinkling increases in spatial complexity with increasing mode number of the cell. In addition, the amplitudes of the wrinkles increase with increasing κ, consistent with the behavior shown in figures 6 and 7 and the generally destabilizing effect of increasing κ found in the analysis of [25]. We point out that as κ increases the curvature of the flame front decreases. Other examples of a destabilizing effect of decreasing curvature have been found for other problems in combustion (for example, [23]).

We next consider the spatial behavior of a specific cellular solution. The analysis for the flame front model demonstrated that, along the flame front, the harmonic would combine with the fundamental in a manner that would produce a pattern of peaks and troughs. The peaks point in the direction of the burned region, i.e. toward the products of combustion, while the relatively flat troughs point toward the cold region, i.e. toward the fuel. This behavior is also present in experimentally observed cellular flames [24].

The computations exhibit a similar behavior. In figures 11a-e we plot temperature and concentration as a function of the polar angle ϕ for various r locations. The results are for a six-cell with $\kappa = 15.6$. The figures show a large sinusoidal oscillation just ahead of the reaction zone. The characteristic crests and troughs of cellular flames appear only in a localized region where the overall oscillation is considerably reduced. This is consistent with the analysis in [25] which predicted a faster decay rate away from the front for the harmonics.

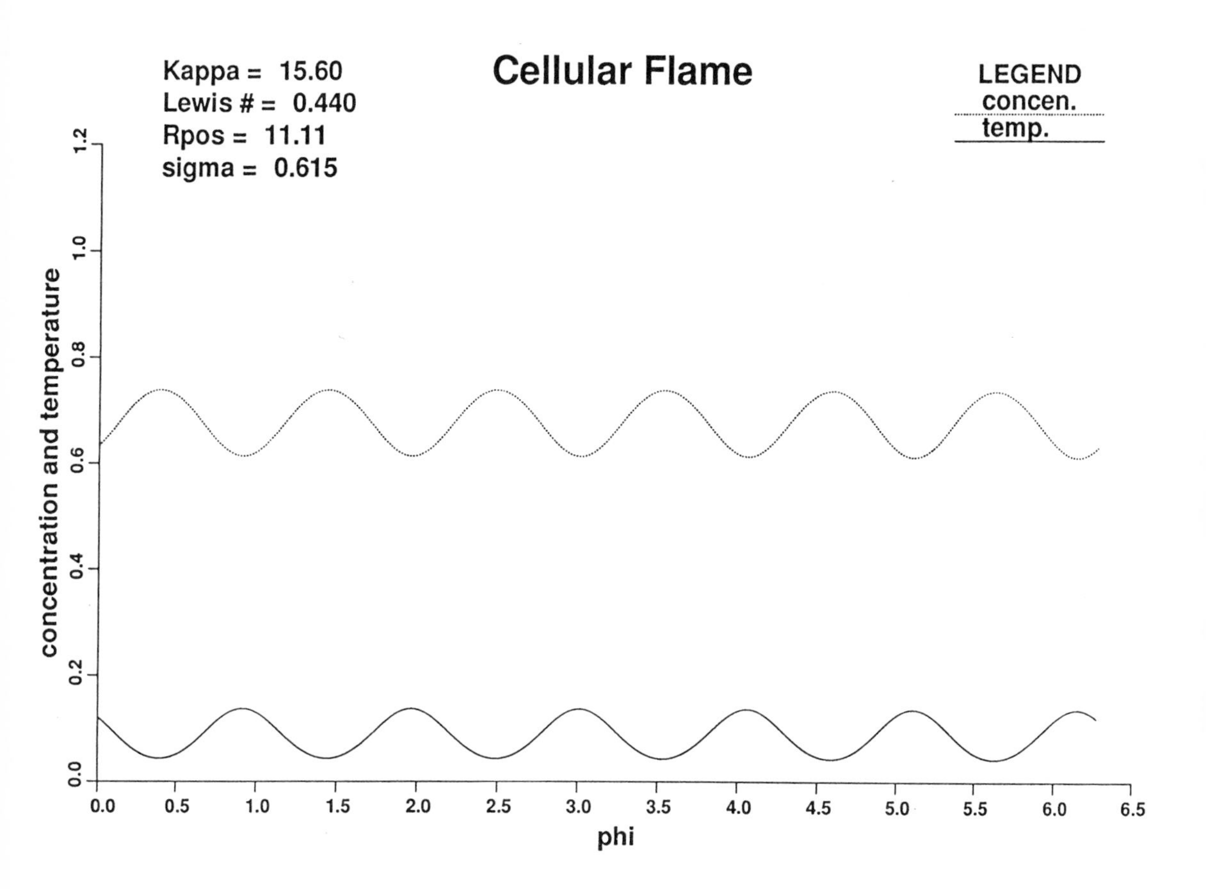

Figure 11(a) Θ and C as a function of ϕ. Six-cell solution, κ = 15.6, r = 11.11.

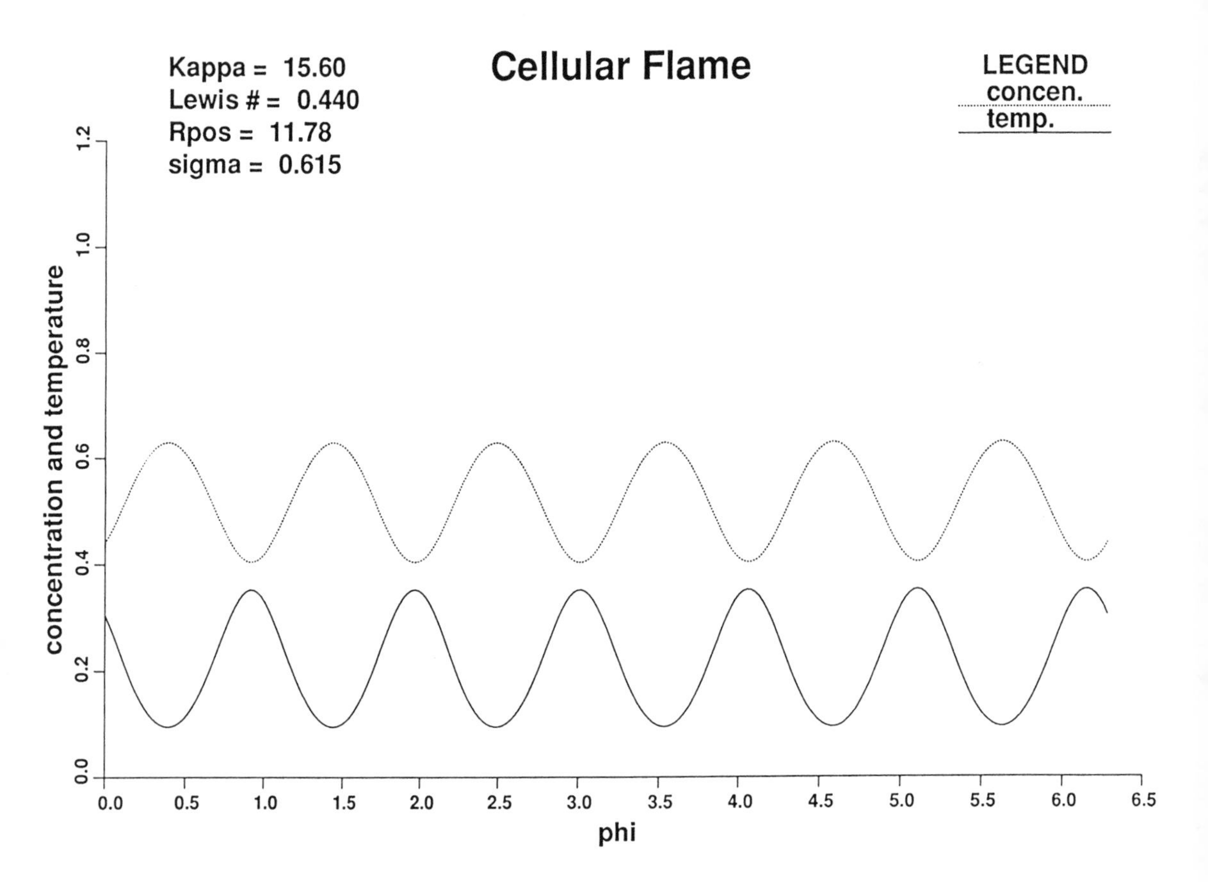

Figure 11(b) Θ and C as a function of ϕ. Six-cell solution, $\kappa = 15.6$, $r = 11.78$.

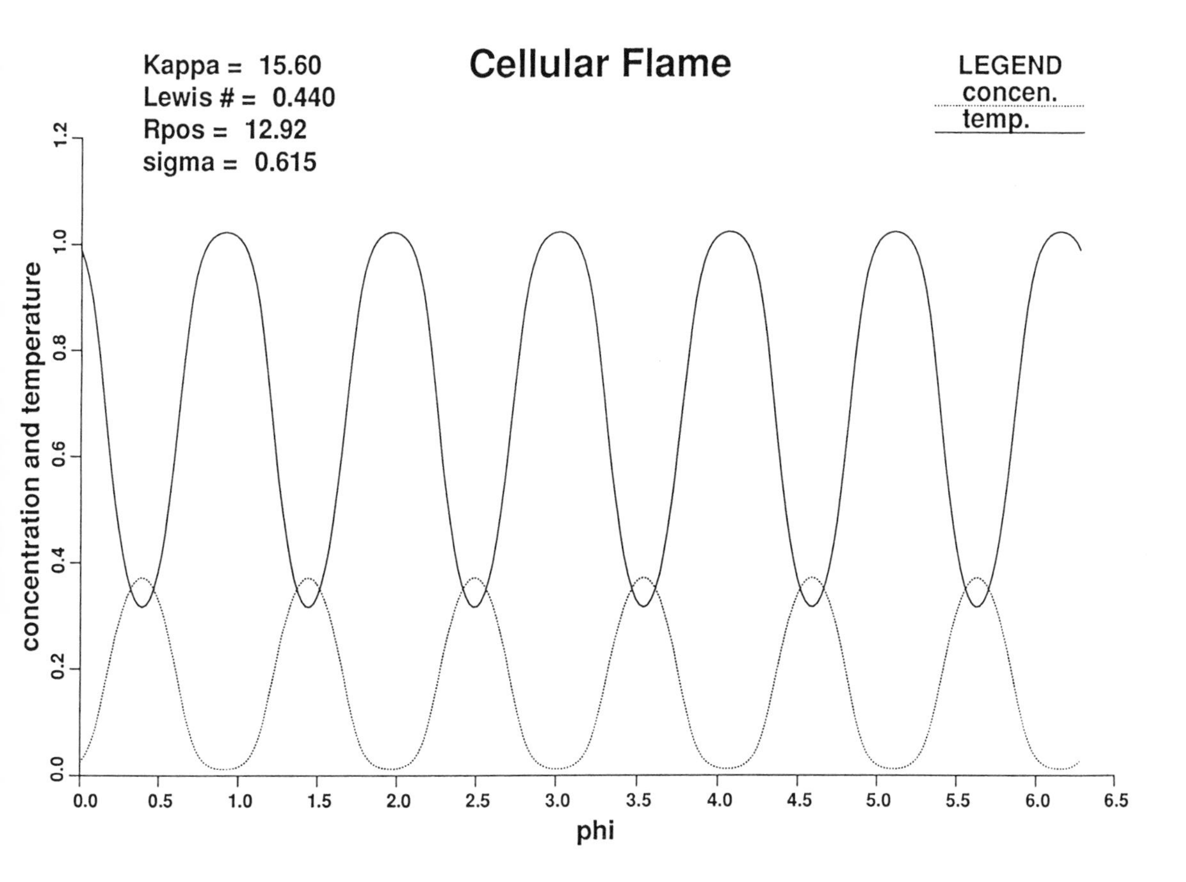

Figure 11(c) Θ and C as a function of ϕ. Six-cell solution, κ = 15.6, r = 12.92.

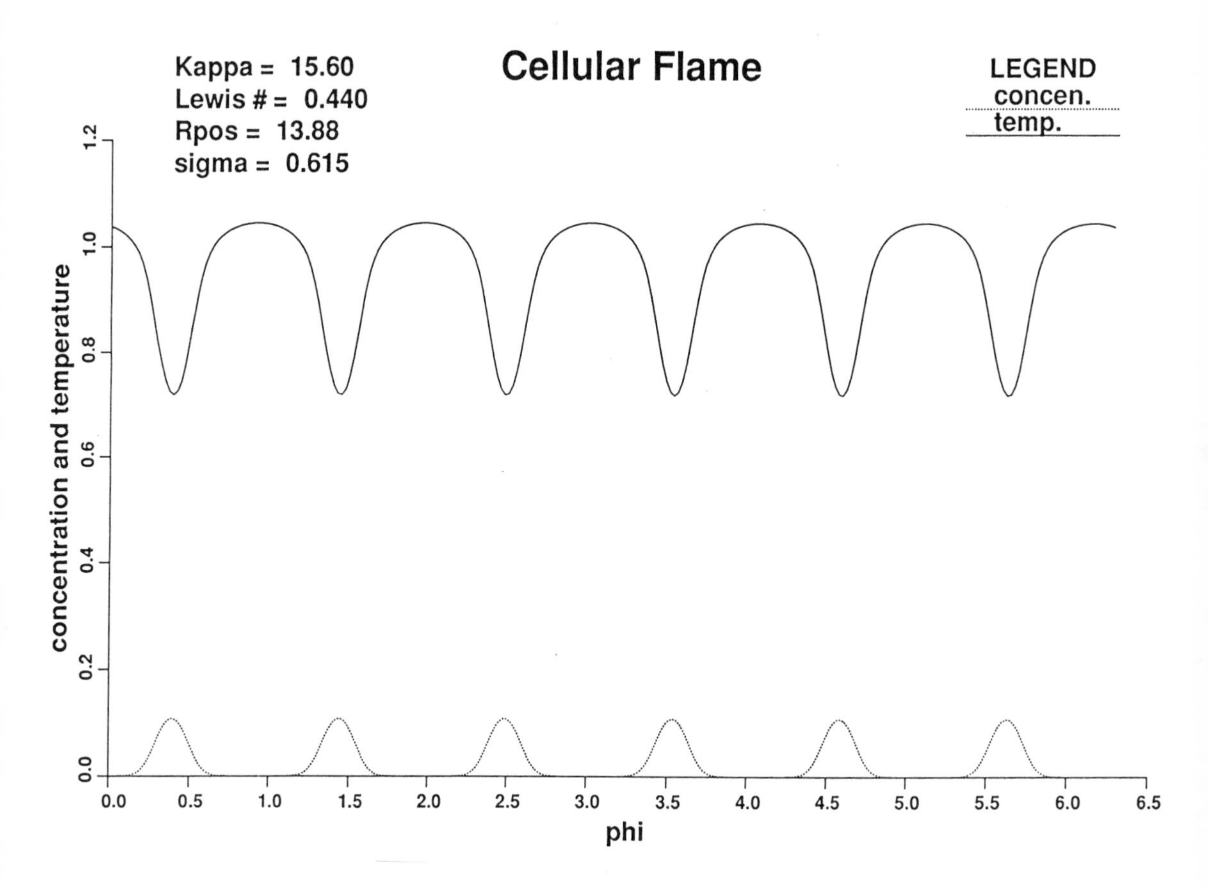

Figure 11(d) Θ and C as a function of ϕ. Six-cell solution, $\kappa = 15.6$, $r = 13.88$.

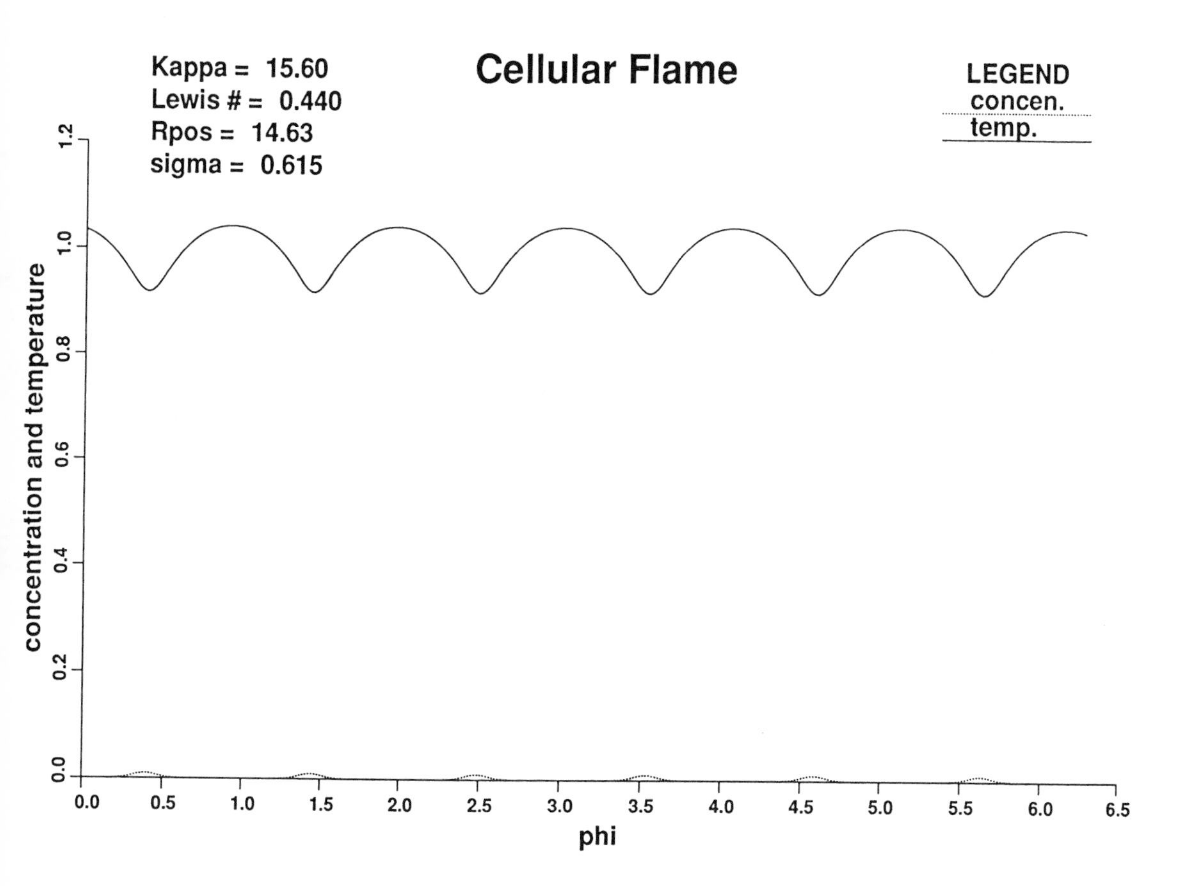

Figure 11(e) Θ and C as a function of ϕ. Six-cell solution, κ = 15.6, r = 14.63.

Lastly we consider the conclusions that can be inferred about the nature of the transitions. In view of the fact that the stationary cells are computed from a time marching algorithm, only stable solutions can be computed. The convergence to a stationary solution can be very deceptive. In figure 12 we plot the logarithm of the residual of the equation for Θ for the case of a transition between a four-cell and a three-cell at $\kappa = 9.25$. The initial data were a four-cell at a larger value of κ. The horizontal axis is the

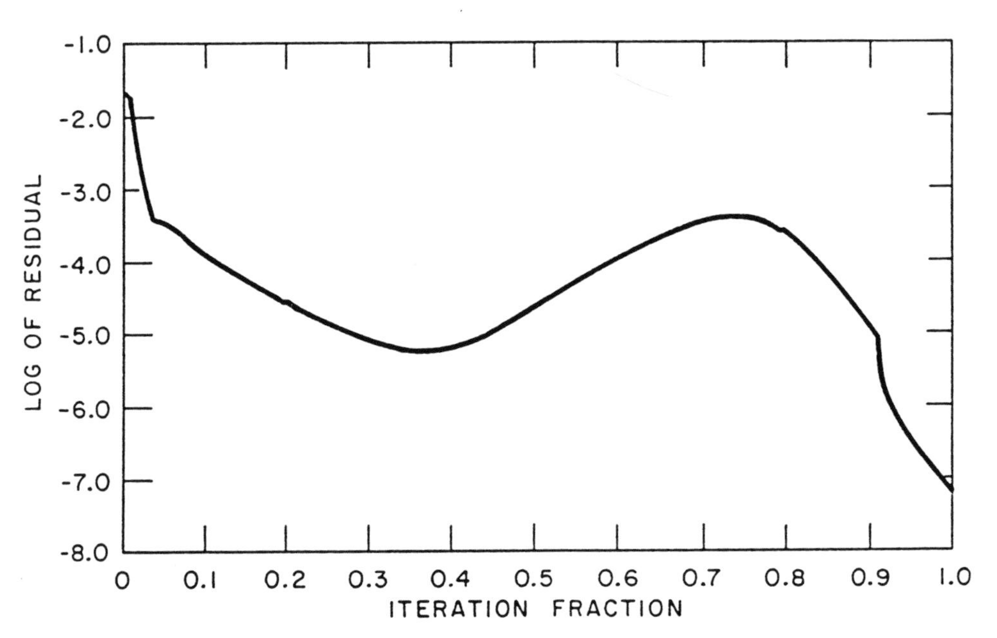

Figure 12 Convergence history for the transition from a four-cell to a three-cell at $\kappa = 9.25$.

fraction of the number of time steps. The time step changed at several times due to the adaptive procedure. The solution appears to converge (to a four-cell) and then begins a slow divergence while the three-mode grows. Finally, the residual decays as the solution converges to a stable, stationary three-cell.

The nature of the transition is clarified by noting that modal solutions can be computed beyond the transition points provided the computational domain is restricted to exclude the unstable angular perturbation. Specifically, we restricted the computation to the sector $0 \leq \phi \leq 2\pi/3$ and enforced periodicity at $\phi = 0$ and $\phi = 2\pi/3$. In this way we were able to compute three-cell solutions beyond the transition point κ_2, at $\kappa = 11.70$ and 11.85. These points are indicated on figures 6 and 7 and demonstrate the continuous extension of the solution branch beyond the point at which it loses stability. Similarly we were able to compute four-cell solutions for values of κ below the value κ^1 where the four-cell loses stability. Although we have not done this for the other solution branches, we believe that the behavior is similar. Since the solution branches exist beyond the points at which they lose stability, it appears that the modal transitions occur via subcritical bifurcations connecting the different modal solution branches.

REFERENCES

[1] A.P. Aldushin, T.M. Martem'yanova, A.G. Merzhanov, B.I. Khaikin and K.G. Shkadinskii, Autovibrational propagation of the combustion front in heterogeneous condensed media, Fizika Goreniya i Vzryva, 9 (1973), 613-626.

[2] A. Bayliss, D. Gottlieb, B.J. Matkowsky, and M. Minkoff, An adaptive pseudo-spectral method for reaction diffusion problems, J. Comput. Phys., 81 (1989), 421-443.

[3] A. Bayliss and B.J. Matkowsky, Fronts, relaxation oscillations, and period doubling in solid fuel combustion, J. Comput. Phys., 71 (1987), 147-168.

[4] D. Gottlieb and S.A. Orszag, C.B.M.S.-N.S.F. Regional Conference Series in Applied Mathematics, No. 26, SIAM, Philadelphia, 1977.

[5] A.P. Hardt and P.V. Phung, Propagation of gasless reactions in solids - 1. Analytical study of exothermic intermetallic reaction rates, Combust. Flame, 21 (1973), 77-89.

[6] J.B. Holt, The use of exothermic reactions in the synthesis and densification of ceramic materials, Mater. Res. Soc. Bull., 12 (1982), 60-64.

[7] B. Larrouturou, Utilisation de maillage adaptifes pour la simulation de flammes monodimeinsionelles instationnaires in Numerical Simulation of Combustion Phenomena, Lecture Notes in Physics, No. 241, Springer-Verlag, Berlin, 1985, pp. 300-312.

[8] B. Larrouturou, Adaptive numerical methods for unsteady flame propagation in Reacting Flows: Combustion and Chemical Reactors, Lectures in Applied Mathematics, No. 24, 1986, pp. 415-435.

[9] Y.A. Maksimov, A.T. Pak, G.B. Lavrenchuk, Y.S. Naiborodenko and A.G. Merzhanov, Spin combustion of gasless systems, Combust. Explos. Shock Waves, 15 (1979), 415-418.

[10] S.B. Margolis, An asymptotic theory of condensed two-phase flame propagation, SIAM J. Appl. Math., 43 (1983), 331-369.

[11] S.B. Margolis, H.G. Kaper, G.K. Leaf, and B.J. Matkowsky, Bifurcation of pulsating and spinning reaction fronts in condensed two-phase combustion, Combust. Sci. Tech., 43 (1985), 127-165.

[12] B.J. Matkowsky and E. Reiss, Singular perturbations of bifurcations, SIAM J. Appl. Math., 33 (1977), 230-255.

[13] B.J. Matkowsky and G.I. Sivashinsky, Propagation of a pulsating reaction front in solid fueld combustion, SIAM J. Appl. Math., 35 (1978), 465-478.

[14] A.G. Merzhanov, SHS processes: Combustion theory and practice, Arch. Combust., 1 (1981), 23-48.

[15] A.G. Merzhanov, A.K. Filonenko and I.P. Borovinskaya, New phenomena in combustion of condensed systems, Dokl. Phys. Chem., 208 (1973), 122-125.

[16] K.G. Shkadinskii, B.I. Khaikin and A.G. Merzhanov, Propagation of a pulsating exothermic reaction front in the condensed phase, Combust., Explos. Shock Waves, 7 (1971), 15-22.

[17] G.I. Sivashinsky, On spinning propagation of combustion waves, SIAM J. Appl. Math. 40 (1981), 432-438.

[18] M.D. Smooke and M.L. Koszykowski, Two dimensional fully adaptive solutions of solid-solid alloying reactions, J. Comput. Phys., 62 (1986), 1-25.

[19] A. Bayliss, B.J. Matkowsky and M. Minkoff, Period doubling gained, period doubling lost, SIAM J. Appl. Math., to appear.

[20] A. Bayliss, B.J. Matkowsky and M. Minkoff, Cascading cellular flames, SIAM J. Appl. Math., to appear.

[21] A. Bayliss, B.J. Matkowsky, and M. Minkoff, Adaptive pseudo-spectral computation of cellular flames stabilized by a point source, Appl. Math. Lett., 1 (1988), 19-24.

[22] A. Bayliss, B.J. Matkowsky, and M. Minkoff, Bistable cellular flames, Proc. Symp. Honoring C. C. Lin, World Scientific, Singapore, (1988).

[23] S.B. Margolis and B.J. Matkowsky, Nonlinear stability and bifurcation in the transition from laminar to turbulent flame propagation, Combust. Sci. Tech., 34 (1983), 45-77.

[24] G.H. Markstein (ed.), Nonsteady Flame Propagation, Pergamon Press, Elmsford, NY, 1967.

[25] B.J. Matkowsky, L.J. Putnick, and G.J. Sivashinsky, A nonlinear theory of cellular flames, SIAM J. Appl. Math., 38 (1980), 489-504.

[26] B.J. Matkowsky and G.I. Sivashinsky, An asymptotic derivation of two models in flame theory associated with the constant density approximation, SIAM J. Appl. Math., 37 (1979), 686-699.

A. Bayliss, B.J. Matkowsky
Department of Engineering Sciences
and Applied Mathematics
The Technological Institute
Northwestern University
Evanston, Illinois 60201,
U.S.A.

and

M. Minkoff
Mathematics and Computer Science
Division
Argonne National Laboratory
Argonne, Illinois 60439,
U.S.A.

P.J. BROWNE

Two-parameter eigencurve theory

This lecture presents a survey of some recent results in two-parameter spectral theory developed by P.A. Binding and the author. Examples of eigenvalue problems in ordinary differential equations with two spectral parameters exist in the classical literature – the Mathieu equation is a well-studied case which readily comes to mind. The first part of the lecture contains an overview of properties of eigencurves in abstract problems, while the second part shows the realization of these theories when applied to second-order ordinary differential equations.

1. THE ABSTRACT PROBLEM

Let H be a separable Hilbert space and in it consider operators T, R, S as follows:

(i) $T:D(T) \subset H \to H$: self-adjoint, bounded below and with compact resolvent,

(ii) $R,S:H \to H$: self-adjoint, bounded.

The eigenvalue problem under consideration is that of finding $(\lambda,\mu) \in \mathbf{R}^2$ so that there is a nontrivial solution $y \in H$ of

$$(T - \lambda S + \mu R)y = 0.$$

There is no *a priori* guarantee that eigenvalues (λ,μ) exist, so we assume R (or some linear combination of R and S) to be strictly positive definite. It then is possible to assume without loss of generality that R = I and so the eigenvalue problem becomes that of finding $(\lambda,\mu) \in \mathbf{R}^2$ so that there is $y \neq 0$ with

$$(T - \lambda S + \mu I)y = 0.$$

---

* Research supported in part by a grant from the NSERC of Canada.

Equivalently, given λ we seek μ so that $\mu \in \sigma(\lambda S - T)$. For any $\lambda \in \mathbb{R}$, $\sigma(\lambda S - T)$ is nonempty and consists of a sequence of eigenvalues $\mu^n(\lambda)$ of finite multiplicity accumulating only at $-\infty$. We index these eigenvalues variationally and thus have

$$\mu^0(\lambda) \geqq \mu'(\lambda) \geqq \mu^2(\lambda) \geqq \dots . \qquad (1)$$

These are the so-called *eigencurves* we wish to study and this lecture details some of their basic properties.

The $\mu^n(\lambda)$ are piecewise analytic: According to Kato's analytic perturbation theory, $\lambda S - T$ has, for varying λ, a sequence $\mu_a^n(\lambda)$ of analytic eigenvalues with a corresponding set of orthonormal eigenvectors $y_a^n(\lambda)$ also analytic in λ. These curves may not have the desired ordering (1) but if we take the pointwise maximum and then the maximum but one and so on, we generate our $\mu^n(\lambda)$ with the desired ordering. The points where our $\mu^n(\lambda)$ fail to be analytic are the points where two or more of the μ_a^n cross. At points of analyticity

$$\frac{d}{d\lambda} \mu^n(\lambda) = (Sy_a^n(\lambda), y_a^n(\lambda))$$

with corresponding formulae for right- and left-hand derivatives holding for all values of λ.

$\mu^0(\lambda)$ is convex: I.e., $\{(\lambda,\mu) \mid \lambda \in \mathbb{R}, \mu \geqq \mu^0(\lambda)\}$ is convex.

DEFINITION: λ is a critical point and $(\lambda,\mu^i(\lambda))$ a critical pair for μ^i if

$$(\mu^i)'_+(\lambda)(\mu^i)'_-(\lambda) \leqq 0,$$

and is *degenerate* if $(\mu^i)' = 0$ is some neighbourhood of λ.

The nondegenerate critical points for μ^i have no finite accumulation.

DEFINITION: The form domain D(t) of T. Select $\alpha \in R$ large enough so that $P = (T + \alpha)^{1/2}$ exists. Define $D(t) = D(P)$ and, for $x \in D(t)$,

$$t(x) = \|Px\|^2 - \alpha.$$

Alternatively, if $T = \int sdE(s)$, where E is the resolution of the identity for T, then

$$D(t) = \{x \mid \int sd(E(s)x,x) < \infty\},$$

$$t(x) = \int sd(E(s)x,x).$$

Note that $t(x)$ is an extension of (Tx,x), $x \in D(T)$.

DEFINITION: $D := N(S) \cap D(t)$, where $N(S)$ is the null space of S.

Condition N is said to hold if $D = \{0\}$.

If N holds there are no degenerate critical points, and if γ is real, the number of critical points (λ,μ) with $\mu \geqq \gamma$ is finite.

ASYMPTOTIC PROPERTIES

DEFINITION: μ^n has $(1,\alpha) \in R^2$ as an *asymptotic direction* $(1,\alpha)$ as $\lambda \to \infty$ if α is an accumulation point of $\mu^n(\lambda)/\lambda$ as $\lambda \to \infty$.

If μ^n has an asymptote of slope α, then $(1,\alpha)$ is the asymptotic direction for μ^n, but μ^n may have an asymptotic direction without having an asymptote.

DEFINITION: If E is the resolution of the identity for S (i.e. $S = \int sdE(s)$), define

$$\alpha_+^i = \inf \{\gamma \mid \dim E(\gamma,\infty)H \leqq i\}.$$

For example, if S has numerical range [a,b] and b is a point of continuous spectrum then $\alpha_+^i = b$ for all i, while if b is an isolated eigenvalue of multiplicity k, then $\alpha_+^i = b$ for $i = 0,\dots,k-1$.

(i) μ^j has $(1,\alpha_+^i)$ as its (unique) asymptotic direction as $\lambda \to \infty$.

We now investigate the problem of characterizing the case in which μ^n has an asymptote as well as that of finding the equation of the asymptote.

It is convenient to use new (skew) coordinates $(\tilde{\lambda},\tilde{\mu})$ related to (λ,μ) via

$$\begin{bmatrix} \tilde{\lambda} \\ \tilde{\mu} \end{bmatrix} = \begin{bmatrix} \sec\theta & 0 \\ -\tan\theta & 1 \end{bmatrix} \begin{bmatrix} \lambda \\ \mu \end{bmatrix},$$

which corresponds to rotating the λ-axis through an angle θ. Then

$$T - \lambda S + \mu I = T - \tilde{\lambda} S(\theta) + \tilde{\mu} I$$

where

$$S(\theta) = (\cos\theta)S - (\sin\theta)I.$$

We intend using $\theta = \tan^{-1}(\alpha_+^i)$. We then see that without loss of generality we can return to the case

$$(T - \lambda S + \mu I)y = 0, \quad \alpha_+^i = 0.$$

Thus the problem now is to find conditions under which

$$\mu^i(\lambda) \to \text{constant as } \lambda \to \infty$$

and to determine the constant.

With $S = \int s\,dE(s)$ we put $N_+(S) = E(0,\infty)H$, $N(S)$ = null space of S, $N_-(S) = E(-\infty,0)H$, $D_+ = N_+(S) \cap D(t)$, $\dim N_+ = d_+$, $D = N(S) \cap D(t)$. The conditions we require are given in terms of the dimensions of these spaces.

(ii) If $\dim N_+(S) > i$ then $\mu^i(\lambda) \to \infty$ as $\lambda \to \infty$. (This is true for general values of θ but with $\alpha_+^i = 0$ we have $\dim N_+(S) \leqq i$.)

(iii) If $\dim N_+(S) \leqq i$ and $\dim (D \oplus N_+) \leqq i$, then $\mu^i(\lambda) \to -\infty$ as $\lambda \to \infty$.

(iv) If $\dim N_+(S) \leqq i$ and $\dim (D \oplus N_+) > i$ then $\mu^i(\lambda) \to$ constant as $\lambda \to \infty$.

Thus result (iv) characterizes the case in which an asymptote exists.

DEFINITION:

$$\tau^j = \text{Min}\ \{\text{Max}\ \{t(u) \mid \|u\| = 1,\ u \in F\} \mid F \subset D,\ \dim F = j + 1\}$$

$$= \text{Max}\ \{\text{Min}\ \{t(u) \mid \|u\| = 1,\ u \in E^{\perp} \cap D\} \mid \dim E = j,\ E^{\perp} \cap D \neq \{0\}\}.$$

These quantities are defined for $j = 0,\ldots,\dim D-1$. They are "stationary" values of t restricted to D.

If $\dim N_+(S) \leqq i$ and $\dim (D \oplus N_+) > i$, then

$$\mu^i(\lambda) \to -\tau^{i-d_+} \text{ as } \lambda \to \infty .$$

This gives the equation of the asymptote.

In terms of the original (λ,μ) coordinates we require S to have α_+^i as an eigenvalue with eigenspace E_+^i. The asymptote has equation

$$\mu = \alpha_+^i\lambda - \tau^{i-d_+}(E_+^i).$$

2. ORDINARY DIFFERENTIAL EQUATIONS OF SECOND ORDER

Here we take $H = L^2([a,b])$,

$$D(T) = \{y \in H \mid y' \in A.C.,\ y'' \in H, \text{ and boundary conditions are satisfied}\}$$

$$Ty = -y'' + qy,$$

where $q \in L^{\infty}([a,b])$ is given. We are also given $s \in L^{\infty}([a,b])$ and define

$$(Sy)(x) = s(x)y(x).$$

The eigenvalue problem is

$$-y'' + qy - \lambda sy + \mu y = 0.$$

Various boundary conditions can be used:

(S) $y(a)\cos\alpha + y'(a)\sin\alpha = 0,\ 0 \leqq \alpha \leqq \pi$,

$y(b)\cos\beta + y'(b)\sin\beta = 0,\ 0 < \beta \leqq \pi$,

(P) $y(a) = y(b),\ y'(a) = y'(b)$,

(A-P) $y(a) = -y(b),\ y'(a) = -y'(b)$.

With S defined as above we have

$$\alpha_+^0 = \alpha = \operatorname*{ess\,sup}_{x \in [a,b]} s(x), \quad \forall i.$$

All eigencurves μ^i have the same asymptotic direction $(1,\alpha)$ and further, either *all* μ^i have asymptotes or *none* do.

μ^i has an asymptote if, and only if, s achieves its essential supremum on some interval(s). The key set is the "essential interior" of $s^{-1}(\alpha)$ given by $\Omega = \{x \mid s(\omega) = \alpha$ for almost all ω in some neighbourhood of $x\}$.

This set is open and can be written as a union of disjoint intervals

$$\Omega = \bigcup_i (a_i, b_i).$$

A characterization of the form domain D(t) for T has been given by Hinton [3] as well as a formula for the quadratic form $t(y)$, $y \in D(t)$. This quadratic form consists of the usual Dirichlet integral plus a term arising from the boundary conditions. We use standard variational theory applied to the minimax process for finding the quantities τ^j. It turns out that to calculate the τ^j one solves the eigenvalue problem

$$-y'' + qy = \tau y$$

on the intervals (a_i, b_i) in Ω subject to boundary conditions as follows:

(a) We use a Dirichlet condition at any $a_i, b_i \neq a, b$, and in addition

(b) we retain (S) if T had been given via (S).

(c) We use $y(a) = y(b)$ if T had been given via (P).

(d) We use $y(a) = -y(b)$ if T had been given via (A-P).

EXAMPLE: Consider Hill's equation

$$-y'' - \lambda sy + \mu y = 0 \quad \text{on } [-1,1]$$

with periodic boundary conditions $y(-1) = y(1)$, $y'(-1) = y'(1)$. We take $s(x)$ as

$$s(x) = 0 \text{ on } [-1,-1/2] \cup [1/2,1]$$

$$s(x) = 1 \text{ on } (-1/2,1/2).$$

Clearly $\alpha = 1$ and $\Omega = (-1/2,1/2)$. Thus the eigencurves have asymptotes and we calculate the τ^i by

$$-y'' = \tau y \text{ on } [-1/2,1/2]$$

subject to Dirichlet conditions to give

$$\tau^i = (j+1)^2\pi^2, \; j = 0,1,2,\ldots .$$

Thus the asymptotes for the eigencurves are

$$\mu = \lambda - (j+1)^2\pi^2.$$

Other examples can be given.

Details of these and further results can be found in [1,2].

REFERENCES

[1] P.A. Binding and P.J. Browne, Spectral properties of two parameter eigenvalue problems II, Proc. R. Soc. Edinburgh, 106A (1987), 39-51.

[2] P.A. Binding and P.J. Browne, Eigencurves for two parameter self-adjoint ordinary differential equations of even order, submitted.

[3] D.B. Hinton, On the eigenfunction expansions of singular ordinary differential equations, J. Differential Equations, 24 (1977), 282-305.

Patrick J. Browne
Department of Mathematics
and Statistics
University of Calgary
Calgary, Alberta,
Canada, T2N 1N4

J. FLECKINGER-PELLÉ

On eigenvalue problems associated with fractal domains

It has been known for more than two centuries (Euler, Lagrange) that the vibrations of a membrane Ω are described by the wave equation

$$\partial_t^2 u(x,t) = c\Delta u(x,t), \quad (x,t) \in \Omega \times \mathbb{R}_+,$$

with some limit conditions. Therefore, the determination of the characteristic frequencies of Ω leads to the following eigenvalue problem:

$$(P_0) \quad \begin{cases} -\Delta u = \lambda u \text{ in } \Omega \\ u = 0 \quad \text{on } \partial\Omega. \end{cases}$$

Problem (P_0) arises also if we study the evolution of temperature in a medium, described by the heat equation.

When Ω is a bounded open set in $\mathbb{R}^n$, $n \geq 1$, (P_0) has an infinite number of eigenvalues:

$$0 < \lambda_1 < \lambda_2 \leqq \lambda_3 \leqq \ldots \leqq \lambda_j \leqq \lambda_{j+1} \leqq \ldots, \; \lambda_j \xrightarrow[j \to \infty]{} +\infty.$$

Here, each eigenvalue is repeated according to its (algebraic) multiplicity.

An analogous result holds for the Neumann problem:

$$(P_1) \quad \begin{cases} -\Delta u = \lambda u \text{ in } \Omega \\ \frac{\partial}{\partial n} u = 0 \text{ on } \partial\Omega. \end{cases}$$

Problems (P_0) and (P_1) are considered in their variational form; in other words, we say that λ is an eigenvalue of (P_0) (resp. (P_1)) if there exists a nonzero u in $H^1(\Omega)$ (resp. $H_0^1(\Omega)$) satisfying $-\Delta u = \lambda u$ in the distributional sense.

Here $H^1(\Omega)$ is the usual Sobolev space of order 1, and $H_0^1(\Omega)$ is the completion of $C_0^\infty(\Omega)$ with respect to the norm of $H^1(\Omega)$.

1. WEYL'S FORMULA

We are interested in the asymptotics of λ_j as j tends to $+\infty$, or what is equivalent, in the asymptotics, as λ tends to $+\infty$, of the "counting function" $N(\lambda)$, also denoted more precisely by $N_i(\lambda,-\Delta,\Omega)$ (i = 0 or 1 according to the boundary conditions), the number of eigenvalues of (P_i) less than λ:

$$N(\lambda) = N_i(\lambda,-\Delta,\Omega) = \#\{\lambda_j \leqq \lambda\} .$$

This problem was introduced by Lorentz and Jeans in the study of electromagnetic radiation theory.

It was proved in Weyl (1911) that, for $\partial\Omega$ smooth enough,

$$N_i(\lambda,-\Delta,\Omega) \sim (2\pi)^{-n}\omega_n|\Omega|_n\lambda^{n/2}, \text{ as } \lambda \to +\infty, \tag{1}$$

where ω_n denotes the volume of the unit ball in $\mathbb{R}^n$, and $|\cdot|_n$ stands for the n-dimensional Lebesgue measure.

1.1 A COUNTEREXAMPLE: Indeed, it was proved in Métivier (1976, 1977) that (1) always holds for the Dirichlet problem (P_0) when Ω is bounded. For the Neumann problem, (1) holds only when the boundary $\partial\Omega$ is not "too long"; it holds in particular if Ω satisfies the "strong cone property". Moreover, the necessity of such a condition is established by the following counterexample (Fleckinger and Metivier, 1973):

For a given positive number s, let us define

$$\Omega_s = \{(x,y) \in \mathbb{R}^2 \mid x \in (0,1);\ 0 < y < 1 + \sum_{j\in\mathbb{N}} j^{-s}\,\mathbf{1}_{I_j}(x)$$

where $(I_j)_{j\in\mathbb{N}}$ is an infinite sequence of disjoint open intervals in (0,1) and where $\mathbf{1}_{I_j}$ denotes the defining function of the set I_j:

$$\mathbf{1}_{I_j}(x) = \begin{cases} 1 & \text{if } x \in I_j \\ 0 & \text{if } x \notin I_j \end{cases} .$$

For this set, when $0 < s < 1/2$, we have

$$N_1(\lambda,-\Delta,\Omega_s) \approx \lambda^{1/2s}, \text{ as } \lambda \to +\infty.$$

The symbol $\approx$ means that there exist two positive constants c' and c" such that

$$c'\lambda^{1/2s} \leqq N_1(\lambda,-\Delta,\Omega_s) \leqq c"\lambda^{1/2s}, \text{ for all } \lambda \text{ large enough.}$$

Therefore, in this example, Weyl's formula does not hold and $\lambda^{-1}N(\lambda) \to +\infty$ as $\lambda \to +\infty$.

1.2. A RECTANGLE: When Ω is a rectangle $(0,a) \times (0,b)$, (1) is very easy to establish, since, in this case, the positive eigenvalues of (P_i) are known:

$$\lambda_{p,q} = p^2\pi^2a^{-2} + q^2\pi^2b^{-2}, \text{ with } (p,q) \in N^* \times N^*.$$

Hence, $N_i(\lambda,-\Delta,\Omega)$ is the number of pairs of integers which are inside a quarter ellipse; it is an old result of number theory (Gauss, 1876)that this number is proportional to the area of this quarter ellipse, and, therefore, (1) holds when Ω is a rectangle in $\mathbf{R}^2$.

This can be extended to a hyperrectangle in $\mathbf{R}^n$.

1.3. EXTENSIONS: When Ω is a bounded open set in $\mathbf{R}^n$, it is possible to prove (1) by the "Weyl-Courant method" which consists of cutting $\mathbf{R}^n$ into cubes and approximating Ω by a union of cubes (see e.g. Courant and Hilbert (1953) or Reed and Simon (1978)). Throughout the proof, the following two results are used:

$$N_0(\lambda,-\Delta,\Omega) \geqq N_0(\lambda,-\Delta,\Omega') \text{ if } \Omega \supseteq \Omega'. \qquad (2)$$

If Ω_1 and Ω_2 are two disjoint open sets in Ω, with $\bar{\Omega} = \overline{\Omega_1} \cup \overline{\Omega_2}$, then

$$N_0(\lambda,-\Delta,\Omega_1) + N_0(\lambda,-\Delta,\Omega_2) \leqq N_0(\lambda,-\Delta,\Omega) \leqq N_1(\lambda,-\Delta,\Omega)$$
$$\leqq N_1(\lambda,-\Delta,\Omega_1) + N_1(\lambda,-\Delta,\Omega_2).$$

This formula is known as "Dirichlet-Neumann bracketing" by physicists.

2. INFLUENCE OF THE BOUNDARY AND APPLICATIONS IN INDUSTRY

When (1) is established, two questions arise naturally.

(i) Is it possible to estimate the "remainder term" $N_i(\lambda,-\Delta,\Omega) - \phi(\lambda,\Omega)$ where $\phi(\lambda,\Omega)$ denotes "Weyl's term":

$$\phi(\lambda,\Omega) = (2\pi)^{-n}\omega_n|\Omega|_n\lambda^{n/2}? \tag{4}$$

(ii) Since the volume $|\Omega|_n$ can be deduced from the knowledge of the spectrum, are other geometrical attributes available with this knowledge?

This was summarized in a famous paper (Kac, 1966) entitled: "Can one hear the shape of a drum?" Indeed, it is impossible to determine Ω completely just from knowledge of the eigenvalues since there exist isospectral drums in $\mathbb{R}^n$, $n \geq 4$, which are not isometric (Urakawa, 1982). Nevertheless, the asymptotics of the eigenvalues determine the "length" of the boundary (or more generally its measure, $|\partial\Omega|_{n-1}$), the number of angles, the number of "holes",... (see e.g. McKean and Singer (1967) and Sleeman and Zayed (1983)). Note that the asymptotics of the eigenvalues are derived from the asymptotics of the counting function $N(\lambda)$ as well as from the asympotics of $Z(t)$ as $t \to 0$, where

$$Z(t) = \sum_{j\in\mathbb{N}} e^{-\lambda_j t}.$$

The estimates on $Z(t)$ are derived from the estimates on the heat kernel associated with our problems (P_i).

If $\partial\Omega$ is smooth enough, the following estimate holds (Seeley, 1978; Ivrii, 1980; Hörmander, 1985):

$$N_i(\lambda,-\Delta,\Omega) = \phi(\lambda,\Omega) + \gamma_n|\partial\Omega|_{n-1}\lambda^{(n-1)/2} + o(\lambda^{(n-1)/2}), \text{ as } \lambda \to +\infty, \tag{5}$$

where γ_n is a number which depends only on n and i (i = 0 or 1). This result is usually established by a Fourier transform of the spectrum and by use of the Fourier-integral operators.

This result, which is of course very important from a mathematical point of view, also has important applications in industry. It is obvious that a "body" Ω has the same volume $|\Omega|_n$ when it is cracked, but it then has a more important boundary $\partial\Omega$. Therefore, any variation in the vibratory response of a body - e.g. the driving shaft of a power plant - indicates a variation in the geometry of the body and hence, it is possible, just by sending vibrations, to detect any crack.

Indeed, it is possible to guess, by use of the counterexample exhibited above, that in some problems the boundary can play an important role, even more important in the counting function than Weyl's term. In the following we will be concerned with "fractal boundaries", which are precisely such that $|\partial\Omega|_{n-1} = +\infty$.

3. ASYMPTOTICS OF THE EIGENVALUES OF THE DIRICHLET LAPLACIAN ON AN OPEN SET WITH FRACTAL BOUNDARY

In 1979, the British physicist M. Berry, studying the scattering of waves by "fractals", suggested substituting in the second term of (5) the Hausdorff measure and dimension of the boundary $\partial\Omega$ for its Lebesgue measure and for (n-1) (Berry, 1979, 1980). This measure was introduced in 1919 by Hausdorff in the following way.

For $h \in \mathbb{R}$ and $\varepsilon > 0$, we set

$$H(\partial\Omega,h,\varepsilon) = \inf \sum_{i\in I} r_i^h,$$

where the infimum is taken over the set of coverings of the boundary $\partial\Omega$ by balls $(B_i)_{i\in I}$, with radii $r_i < \varepsilon$.

The function $\varepsilon \to H(\partial\Omega,h,\varepsilon)$ decreases with ε, and the Hausdorff h-dimension of $\partial\Omega$ is

$$H_h(\partial\Omega) = \lim_{\varepsilon\to 0} H(\partial\Omega,h,\varepsilon).$$

Moreover, $h_0 = h_0(\partial\Omega)$ defined by

$$H_h(\partial\Omega) = \begin{cases} +\infty & \text{if } h < h_0 \\ 0 & \text{if } h > h_0 \end{cases}$$

is called the Hausdorff dimension of $\partial\Omega$.

This Hausdorff dimension has been widely studied since (Mandlebrot, 1982) and it seems to be the most popular dimension among mathematicians. Nevertheless, it was proved in Brossard and Carmona (1986), with the help of a counterexample consisting of a union of cubes, that Berry's conjecture may fail with the Hausdorff dimension, and they suggested replacing it by Minkowski's one.

This Minkowski dimension has been introduced in different ways by many authors, and therefore it has different names such as "Cantor-Minkowski dimension", "ordre dimensionnel" (by Bouligand), "Bouligand dimension", "logarithmic density", "box counting dimension", "Kolmogorov entropy",.... These definitions are equivalent (Bouligand, 1928; Tricot, 1981). We give here some of them.

For a given positive number ε, we set

$$\Omega_\varepsilon = \{x \in \mathbb{R}^n \mid d(x,\partial\Omega) < \varepsilon\} \tag{6}$$

where $d(x,\partial\Omega)$ denotes the Euclidian distance of x to the boundary $\partial\Omega$. We then consider the positive numbers d such that

$$\lim_{\varepsilon\to 0} \varepsilon^{-(n-d)} |\Omega_\varepsilon|_n = 0,$$

and we define $\delta := \delta(\partial\Omega)$, the Minkowski dimension of $\partial\Omega$, as

$$\delta = \sup\{d \in R_+ \mid \lim_{\varepsilon\to 0} \varepsilon^{-(n-d)} |\Omega_\varepsilon|_n = 0\}. \tag{7}$$

We also have (Bouligand, 1928)

$$\delta = \inf\{d > 0 \mid \limsup_{\varepsilon\to 0} \varepsilon^d n(\varepsilon,\partial\Omega) = 0\},$$

where $n(\varepsilon,\partial\Omega)$ denotes the minimal number of balls with radius ε which are necessary to cover $\partial\Omega$. Therefore δ can also be introduced as an "entropy":

$$\delta = \limsup_{\varepsilon\to 0} \frac{\ln n(\varepsilon,\partial\Omega)}{-\ln \varepsilon} .$$

Let us consider a "grid" in $\mathbb{R}^n$ where the cubes $(Q_\zeta)_{\zeta\in\mathbb{Z}^n}$, for a given integer N, are such that

$$Q_\zeta = \prod_{k=1}^{n} (\zeta_k/N, (\zeta_k+1)/N), \text{ with } \zeta_k \in \mathbb{N}.$$

A "maille utile" for Bouligand is a cube Q_ζ which intersects the boundary $\partial\Omega$, and, if $m(N^{-1},\partial\Omega)$ denotes the number of "mailles utiles", we have

$$\delta = \lim_{N\to+\infty} \frac{\ln m(N^{-1},\partial\Omega)}{\ln N} .$$

Equivalently, in computer science, the "pixels" are defined by $A_\zeta = (\zeta_k/N)_{k=1,\dots,n} \in \mathbb{R}^n$, when the associate cubes Q_ζ intersect the boundary.

Other definitions of the Minkowski dimension are possible (see e.g. Tricot, 1981). We also mention the following inequality which was used in Brossard and Carmona (1986); with the above notations, we have

$$h_0(\partial\Omega) \leq \delta(\partial\Omega).$$

In Lapidus and Fleckinger-Pellé (1987, 1988), the following result is established:

<u>THEOREM 1</u>: If Ω is bounded and if $\partial\Omega$ is fractal, with Minkowski dimension $\delta \in (n-1,n)$, we have

$$N_0(\lambda,-\Delta,\Omega) = (2\pi)^{-n}\omega_n|\Omega|_n\lambda^{n/2} + O(\lambda^{\delta/2}), \text{ as } \lambda \to +\infty.$$

This result has been extended in Lapidus (1988) to more general elliptic operators and to Neumann boundary value problems.

Here we shall prove:

<u>THEOREM 2</u>: If Ω is bounded and if $\partial\Omega$ is δ-Minkowski measurable, with δ-Minkowski measure μ satisfying

$$\lim_{\varepsilon\to 0} \varepsilon^{-(n-\delta)} |\Omega|_n = \mu, \tag{8}$$

then there exist two positive numbers λ_0 and $c(n,\delta)$ such that

$$|N_0(\lambda,-\Delta,\Omega) - \phi(\lambda,\Omega)| \leq c(n,\delta)\mu\lambda^{\delta/2}, \text{ for all } \lambda \geq \lambda_0. \tag{9}$$

In the following we will introduce a positive number ε_0 such that, for all $\varepsilon \in (0,\varepsilon_0)$,

$$\varepsilon^{-(n-\delta)}|\Omega_\varepsilon|_n \leqq 2\mu. \qquad (10)$$

Then, we choose $p_0 \in \mathbb{N}$ such that

$$2^{-p_0} \leqq \varepsilon_0. \qquad (11)$$

4. PROOF

We proceed as in Métivier (1976, pp. 36-37), Métivier (1977, pp. 197-199) or Lapidus and Fleckinger (1987).

For each integer p, we consider as in Courant and Hilbert (1953) a tessalation of $\mathbb{R}^n$ into congruent and non-overlapping cubes $(Q_{\zeta_p})_{\zeta_p \in \mathbb{Z}^n}$ with side $\eta_p = 2^{-(p+p_0)}$. We define by induction

$$A_0 = \{\zeta_0 \in \mathbb{Z}^n \mid Q_{\zeta_0} \subset \Omega\} \quad \text{and } \Omega_0' = \bigcup_{\zeta_0 \in A_0} Q_{\zeta_0};\ \Omega_0'' = \Omega \setminus \overline{\Omega_0'}$$

$$A_1 = \{\zeta_1 \in \mathbb{Z}^n \mid Q_{\zeta_1} \subset \Omega_0''\} \quad \text{and } \Omega_1' = \Omega_0' \cup (\bigcup_{\zeta_1 \in A_1} Q_{\zeta_1});\ \Omega_1'' = \Omega \setminus \overline{\Omega_1'}$$

...

$$A_p = \{\zeta_p \in \mathbb{Z}^n \mid Q_{\zeta_p} \subset \Omega_{p-1}''\} \text{and } \Omega_p' = \Omega_{p-1}' \cup (\bigcup_{\zeta_p \in A_p} Q_{\zeta_p});\ \Omega_p'' = \Omega \setminus \overline{\Omega_p'}.$$

We also define for each integer p

$$B_p = \{\zeta_p \in \mathbb{Z}^n \mid Q_{\zeta_p} \cap \partial\Omega \neq \emptyset;\ Q_{\zeta_p} \cap \Omega_p' = \emptyset\} \text{ and } R_p = \bigcup_{\zeta_p \in B_p} Q_{\zeta_p}.$$

We first make some simple observations.

REMARK 1: $\Omega_p'' \subset \Omega_{\varepsilon_p}$ with $\varepsilon_p = \sqrt{n}\,\eta_p = \sqrt{n}\,2^{-(p+p_0)}$; the set Ω_ε has been defined by (6). Moreover

$$R_p \subset \Omega_{\varepsilon_p}.$$

REMARK 2: Since $\sqrt{n}\,\eta_p \leqq \varepsilon_0$ for all integer p, we deduce from (10) and (11) and from Remark 1 that

$$(\# A_p)\eta_p^n \leqq |\Omega''_{p-1}|_n \leqq |\Omega_{\varepsilon_{p-1}}|_n \leqq 2\mu n^{(n-\delta)/2}(\eta_{p-1})^{n-\delta}.$$

Hence

$$(\# A_p) \leqq \gamma_1\, \mu \eta_p^{-\delta} \tag{12}$$

with

$$\gamma_1 = 2\,(2\sqrt{n})^{(n-\delta)}. \tag{13}$$

An analogous calculation shows that

$$(\# B_p) \leqq \gamma_1 \mu\, \eta_p^{-\delta}. \tag{14}$$

REMARK 3: Since the positive eigenvalues of (P_i) on a cube Q_{ζ_p} with side η_p are $\pi^2\eta_p^{-2}(q_1{}^2 + q_2{}^2 + \ldots + q_n{}^2)$, with $q_j \in \mathbb{N}$, for any integer

$$p \geqq P := \max\{q \in \mathbb{N} \mid \pi^2\eta_q^{-2} < \lambda\} = \max\{q \in \mathbb{N} \mid 2^{(q+p_0)} < \pi^{-1}\lambda^{1/2}\},$$

we have

$$N_i(\lambda, -\Delta, Q_{\zeta_p}) = 0.$$

During the proof, we also make use of the following estimate (Courant and Hilbert, 1953, Section VI.4, or Reed and Simon, 1978, Proposition 2, pp. 266-267).

PROPOSITION 1: There exists a positive constant c', depending only on n, such that, for all cubes $Q \subset \mathbb{R}^n$, with side η, and for all $\lambda > 0$

$$|N_i(\lambda, -\Delta, Q) - \phi(\lambda, Q)| \leqq c'[1 + (\lambda\eta^2)^{(n-1)/2}].$$

It follows from (2), (3) and Remark 3 that

$$\sum_{k=0}^{P} \sum_{\zeta_k \in A_k} N_0(\lambda,-\Delta,Q_{\zeta_k}) \leq N_0(\lambda,-\Delta,\Omega) \leq N_0(\lambda,-\Delta, \overline{\Omega} \hat{\cup} \overline{R_P}$$

$$\leq \sum_{k=0}^{P} \sum_{\zeta_k \in A_k} N_1(\lambda,-\Delta,Q_{\zeta_k}) + \sum_{\zeta_P \in B_P} N_1(\lambda,-\Delta,Q_{\zeta_P}).$$

Then, by subtracting "the Weyl's term" $\phi(\lambda,\Omega)$, we obtain

$$\sum_{k=0}^{P} (\# A_k)[N_0(\lambda,-\Delta,Q_{\zeta_k}) - \phi(\lambda,Q_{\zeta_k})] - [\phi(\lambda,\Omega) - \phi(\lambda,\Omega_P')]$$

$$\leq N_0(\lambda,-\Delta,\Omega) - \phi(\lambda,\Omega)$$

$$\leq \sum_{k=0}^{P} (\# A_k)[N_1(\lambda,-\Delta,Q_{\zeta_k}) - \phi(\lambda,Q_{\zeta_k})] + (\# B_P)[N_1(\lambda,-\Delta,Q_{\zeta_P})-\phi(\lambda,Q_{\zeta_P})]$$

$$+ \phi(\lambda,\Omega_P') + (\# B_P)\ \phi(\lambda,Q_{\zeta_P}) - \phi(\lambda,\Omega). \tag{15}$$

We first examine the "interior term":

$$T_1 = \sum_{k=0}^{P} (\# A_k)\ [N_1(\lambda,-\Delta,Q_{\zeta_k}) - \phi(\lambda,Q_{\zeta_k})).$$

By use of Proposition 1, we obtain

$$|T_1| \leq \sum_{k=0}^{P} (\# A_k)[1 + (\lambda\eta_k^2)^{(n-1)/2}].$$

Then, by Remarks 2 and 3, we have

$$|T_1| \leq \sum_{k=0}^{P} \gamma_1 \mu \eta_k^{-\delta}[1+\lambda^{(n-1)/2}\ \eta_k^{(n-1)}] \leq \gamma_1\ \mu\,(\frac{1}{2^{\delta}-1} + \frac{1}{2^{\delta-(n-1)}-1})\lambda^{\delta/2}. \tag{16}$$

An analogous calculation for the "boundary term"

$$T_2 = (\# B_P)\ [N_1(\lambda,-\Delta,Q_{\zeta_P}) - \phi(\lambda,Q_{\zeta_P})]$$

gives a similar estimate:

$$|T_2| \leq \gamma_1 \mu \lambda^{\delta/2} \, 2^{(p_0+1)} . \tag{17}$$

We consider now

$$T_3 = \phi(\lambda,\Omega) - \phi(\lambda,\Omega_p') = \phi(\lambda,\Omega_p'').$$

By definition of $\phi(\lambda,\Omega)$ (equation 4)), by definition of P (Remark 3) and by Remark 1, it follows easily that

$$|T_3| \leq (2\pi)^{-n} \omega_n \lambda^{n/2} |\Omega_{\varepsilon_p}|_n \leq \mu \gamma_1 \omega_n \lambda^{\delta/2} . \tag{18}$$

For the same reason, we still have

$$T_4 = (\# \, B_p) \phi(\lambda, Q_{\zeta_p}) \leq \mu \gamma_1 \omega_n \lambda^{\delta/2} . \tag{19}$$

Hence, by combining (15) to (19), we obtain Theorem 2. □

ACKNOWLEDGEMENTS: The author would like to thank the British Council and the Science and Engineering Research Council for their financial support.

REFERENCES

Berry, M.V. (1979). Distribution of modes in fractal resonators, in "Structural Stability in Physics", (W. Güttinger and H. Eikemeier, eds.), Springer-Verlag, Berlin, pp. 51-53.

Berry, M.V. (1980). Some geometrical aspects of wave motion: wavefront dislocations, diffractions catastrophes, diffractals, in "Geometry of the Laplace Operator", Proc. Symp. Pure Math., Vol. 36, Amer. Math. Soc., Providence, R.I., pp. 13-38.

Bouligand, G. (1928). Ensembles impropres et nombre dimensionnel, Bull. Sci. Math., 52, (2), 320-344 and 361-376.

Brossard, J. and Carmona, R. (1986). Can one hear the shape of a fractal?, Commun. Math. Phys. 104, 103-122.

Courant, R. and Hilbert, D. (1953). "Methods of Mathematical Physics", Interscience, New York.

Fleckinger, J. (1973). Théorie spectrale des opérateurs uniformément elliptiques sur quelques ouverts irréguliers, in "Séminaires d'Analyse Numérique", Université P. Sabatier, Toulouse, exp.D.

Fleckinger, J. and Métivier, G. (1973). Théorie spectrale des opérateurs uniformément elliptiques sur quelques ouverts irréguliers, C.R. Acad. Sci. Paris, Sér. A, 276, 913-916.

Gauss an Encke (1876). Werke. Göttingen.

Ivrii, V. Ja. (1980). Second term of the spectral asymptotic expansion for the Laplace-Beltrami operator on manifolds with boundary, Funct. Anal. Appl., 14, 98-106.

Kac, M. (1966). Can one hear the shape of a drum?, Amer. Math. Monthly, 73, 1-23.

Lapidus, M.L. (1988). Fractal drum, inverse spectral problems for elliptic operators and a partial solution of the Weyl-Berry conjecture (to appear).

Lapidus, M.L. and Fleckinger-Pellé, J. (1988). Tambour fractal: vers une résolution de la conjecture de Weyl-Berry pour les valeurs propres du Laplacien, C.R. Acad. Sci. Paris, Sér. I, Math. (to appear).

MacKean, H.P. and Singer, I.M. (1967). Curvature and the eigenvalues of the Laplacian. J. Diff. Geom. 1. p. 43-69.

Mandelbrot, B.B. (1982). "The Fractal Geometry of Nature", W.H. Freeman, San Francisco.

Métivier, G. (1973). Théorie spectrale d'opérateurs elliptiques sur des ouverts irréguliers, Sém. Goulaic-Schwartz, Ecole Polytechnique, Paris.

Métivier, G. (1976). Etude asymptotique des valeurs propres de la fonction spectrale de problèmes aux limites, Thèse d'Etat, Université de Nice, France.

Métivier, G. (1977). Valeurs propres de problèmes aux limites elliptiques irréguliers, Bull. Soc. Math. Fr. Mem., 51-52, 125-219.

Reed, M. and Simon, B. (1978). "Methods of Modern Mathematical Physics", Vol. IV. Academic Press, New York.

Seeley, R.A. (1978). A sharp asymptotic remainder estimate for the eigenvalues of the Laplacian in a domain of R^3, Adv. in Math., 29: p.244-269.

Sleeman, B.D. and Zayed, E.M.E. (1983). An inverse eigenvalue problem for a general convex domain, J. Math. Anal. Appl., 94, 78-95.

Tricot, B. (1981). Douze definitions de la densité logarithmique, C.R. Acad. Sci. 23, 549.

Urakawa, H. (1982). Bounded domains which are isospectral but not congruent, Ann. Sci. Ecole Normale Sup., 15: p. 441-456.

Weyl, H. (1912). Das asymptotische Verteilungsgesetz der Eigenwerte linearer partieller Differentialgleichungen, Math. Ann., 71: p. 441-479.

J. Fleckinger-Pellé
LAAS et Université Paul Sabatier
Laboratoire d'Analyse Numérique
118 Route de Narbonne
31062 Toulouse Cedex
France

R. LEIS

Initial-boundary value problems in elasticity

During the last decade much work has been done in elasticity and especially in thermoelasticity. One wants to solve initial-boundary value problems first, and afterwards one tries to get more specific knowledge of the solutions obtained. For instance, one enquires about their regularity, or about their asymptotic behaviour for large times. Meanwhile linear problems are fairly well understood whereas many nonlinear problems are still open.

Initial-boundary value problems play an important role in mathematical physics. We only remind the reader of the wave equation, the Maxwell equations, the Scrhödinger equation, or the system of equations of elasticity. In the linear case all these problems lead to a self-adjoint operator such that the asymptotic behaviour of the solutions and the existence of wave operators can be obtained by means of spectral theory. In linear thermoelasticity, however, the underlying operator is not self-adjoint owing to the coupling of a hyperbolic with a parabolic equation. Although solutions can be obtained with the aid of semigroup theory, it is more difficult - and presumably more interesting - to derive their asymptotic behaviour.

The lecture is organized as follows. We start by treating initial-boundary value problems of linear elasticity, and indicate the existence of wave operators. The main part of the lecture (section 2) is concerned with linear thermoelasticity. We solve initial-boundary value problems, give some asymptotic expansions as $t \to \infty$, and especially describe the asymptotic behaviour of solutions of the free-space problem. In the third section, finally, we deal with some nonlinear questions. We are particularly interested in obtaining solutions global in t for small smooth data.

1. LINEAR ELASTICITY

We start by presenting a brief survey of results obtained in linear elasticity. The underlying equation of state is Hooke's law (stress-strain relation)

$$\tau_{jk} = \sum_{m,n=1}^{3} C_{jkmn} U_{mn}, \quad j,k = 1,2,3. \tag{1.1}$$

τ_{jk} is the stress tensor, $U_{mn} := (\partial_m U_n + \partial_n U_m)/2$ the strain tensor, and $U := (U_1, U_2, U_3)$ the elastic displacement vector. It is defined in an open and connected domain Ω in R^3. The elastic moduli C_{jkmn} are real-valued, bounded and measurable functions on Ω. From physical considerations they display the following symmetry relations

$$C_{jkmn} = C_{mnjk} = C_{kjmn}.$$

Furthermore

$$\exists c_1 > 0 \quad \forall \zeta_{ij} \in \mathbb{C},\ \zeta_{ij} = \zeta_{ji} \quad \forall x \in \Omega \quad \zeta_{jk} C_{jkmn}(x) \bar{\zeta}_{mn} \geq c_1 \sum_{j,k} |\zeta_{jk}|^2$$

holds. For exterior domains Ω (domains with bounded complement) we assume the existence of a sufficiently large constant e such that the C_{jkmn} are constants in $\Omega_e := \{x \in \Omega \mid |x| > e\}$.

I prefer rewriting Hooke's law using Sommerfeld's terminology. Let

$$\alpha_1 = \tau_{11} \quad \alpha_2 = \tau_{22} \quad \alpha_3 = \tau_{33} \quad \alpha_4 = \tau_{23} \quad \alpha_5 = \tau_{31} \quad \alpha_6 = \tau_{12}$$

$$\varepsilon_1 = U_{11} \quad \varepsilon_2 = U_{22} \quad \varepsilon_3 = U_{33} \quad \varepsilon_4 = 2U_{23} \quad \varepsilon_5 = 2U_{31} \quad \varepsilon_6 = 2U_{12}.$$

Using the generalized gradient symbol

$$D := \begin{pmatrix} \partial_1 & 0 & 0 \\ 0 & \partial_2 & 0 \\ 0 & 0 & \partial_3 \\ 0 & \partial_3 & \partial_2 \\ \partial_3 & 0 & \partial_1 \\ \partial_2 & \partial_1 & 0 \end{pmatrix}$$

Hooke's law then reads

$$\alpha = S \cdot DU \tag{1.2}$$

where S is a positive definite six-row matrix containing the elastic moduli, and DU represents the strain tensor.

The elastic medium may show certain symmetries. Special anisotropic media are monoclinic media (one axis of symmetry), rhombic media (two axes), or cubic media (three axes). For details compare Sommerfeld (1949, p. 278f) or Leis (1986, p. 201f). The isotropic medium is given by

$$S = \begin{pmatrix} \nu & \kappa & \kappa & 0 & 0 & 0 \\ \kappa & \nu & \kappa & 0 & 0 & 0 \\ \kappa & \kappa & \nu & 0 & 0 & 0 \\ 0 & 0 & 0 & \mu & 0 & 0 \\ 0 & 0 & 0 & 0 & \mu & 0 \\ 0 & 0 & 0 & 0 & 0 & \mu \end{pmatrix}$$

where μ and κ are the Lamé constants, $\mu > 0$, $2\mu + 3\kappa > 0$, and $\nu := 2\mu + \kappa > 0$.

Using this notation potential and kinetic energy are given by

$$(DU,\alpha)_{L^2(\Omega)} \text{ and } (U_t, MU_t)_{L^2(\Omega)} \tag{1.3}$$

where M is the positive definite density matrix, and for short we write L^2 instead of $(L^2)^6$ or $(L^2)^3$ respectively. Thus the equations of linear elasticity read

$$MU_{tt} - D'SDU = 0. \tag{1.4}$$

$U^0 := U(0)$ and $U^1 := U_t(0)$ are the initial values.

When Ω is equipped with a boundary $\partial\Omega$, we may set boundary value problems. For simplicity we take $M = \mathrm{id}$ and formulate the Dirichlet problem. Let

$$E : \mathcal{D}(E) \subset L^2(\Omega) \to L^2(\Omega)$$

$$U \to -D'SDU$$

where

$$\mathcal{D}(E) := \{U \in \overset{\circ}{H}_1(\Omega) \mid D'SDU \in L^2(\Omega)\}.$$

$(DU, SDU) \geqq p|U|_1^2$ with $p > 0$ then easily follows, and E is a self-adjoint operator (in the case of the Neumann problem one has to use the second Korn's inequality). Thus we can define $E^{\frac{1}{2}}$ with $\mathcal{D}(E^{\frac{1}{2}}) = \overset{\circ}{H}_1$. Assuming $U^0 \in \mathcal{D}(E^{\frac{1}{2}})$ and $U^1 \in L^2$ we obtain a "weak solution with finite energy", $U \in C(\mathbb{R}_0^+, \mathcal{D}(E^{\frac{1}{2}}) \cap C_1(\mathbb{R}_0^+, L^2)$, of

$$U_{tt} + EU = 0 \tag{1.5}$$

through

$$U(t) := \cos(E^{\frac{1}{2}}t)U^0 + E^{-\frac{1}{2}} \sin(E^{\frac{1}{2}}t)U^1, \tag{1.6}$$

which has to be interpreted by the spectral formula for self-adjoint operators. Then

$$\|E^{\frac{1}{2}}U\|^2 + \|U_t\|^2 = \text{const}$$

is the energy.

Let Ω be bounded. Using Rellich's selection theorem one can then easily prove the existence of a countable number of eigenvalues of E, and one obtains a representation of U in terms of standing waves, which is typical for such equations.

More interesting is the case of exterior domains. So let Ω be an exterior domain now. For simplicity we also assume the medium to be isotropic. Only a few anisotropic media have so far been dealt with. In this case a solution U of $(E-\lambda)U = 0$ can be decomposed in Ω_e into a solenoidal and a potential component. Both components solve Helmholtz equations with different wave-numbers. Applying Rellich's estimate and the principle of unique continuation (Weck 1969), one can conclude that E has no point eigenvalues. Here we have to assume that the coefficients C_{ijkm} are differentiable. The spectrum of E is continuous, and $\sigma(E) = \mathbb{R}_0^+$. It is also possible to formulate radiation conditions for each component of U, and to solve exterior boundary value problems

$$(E-\lambda)U = F, \; \lambda \in \mathbb{R}^+$$

for F with finite support ($F \in H^f$). For details compare Leis (1970), or Leis (1986, p. 217). Thus one can prove the limiting absorption principle, which says that one obtains an outgoing or incoming solution (U^+ or U^-) of the exterior boundary value problem by taking the limit in $\mathring{H}_{1,\rho}$ ($\mathring{H}_1$ with weight $\rho(x) := 1/\{1 + |x|\}$)

$$U^{\pm}(\lambda) = \lim_{\varepsilon \downarrow 0} [E - (\lambda \pm i\varepsilon)]^{-1} F.$$

Stone's formula then provides the spectral family of E

$$(P(\lambda)\ U,\ V) = \lim_{\varepsilon \downarrow 0} \frac{1}{2\pi i} \int_0^\lambda ([E - (\mu + i\varepsilon)]^{-1} U - [E - (\mu - i\varepsilon)]^{-1} U,\ V) d\mu,$$

and one concludes that the spectrum of E is absolutely continuous saying that $(P(\lambda)U, U)$ is absolutely continuous for all $U \in L^2$.

This result is strong enough to yield asymptotic statements for U(t) as $t \to \infty$. Using the Riemann-Lebesgue lemma we can easily show "local energy decay" saying that for all r and $\Omega_r := \{x \in \Omega \mid |x| < r\}$

$$\lim_{t\to\infty} \int_{\Omega_r} \{|E^{\frac{1}{2}} U|^2 + |U_t|^2\} = 0. \tag{1.7}$$

Thus one expects that, for large t, U behaves like a free-space solution. Let $C_{0,ijkl} := C_{ijkl}|\Omega_e$ and E_0 be the corresponding operator defined in $\mathbb{R}^3$. It is relatively easy then to discuss the solutions of $\partial_t^2 U_0 + E_0 U_0 = 0$ using the Fourier transform. Knowing that the spectrum of E is absolutely continuous we can apply perturbation methods originated by Kato (1976), Belopolskii and Birman (1968), Pearson (1978) and others. Assuming $U^1 \in \mathcal{D}(E^{-\frac{1}{2}})$ and using complex notation, $H := U^0 + iE^{-\frac{1}{2}} U^1 \in L^2(\Omega)$, we can show that wave operators

$$W^{\pm} : L^2(\Omega) \to L^2(\mathbb{R}^3),\ \text{unitary}$$

$$W^{\pm} := \text{s-}\lim_{t\to\infty} e^{iE_0^{\frac{1}{2}} t}\ J\ e^{-iE^{\frac{1}{2}} t}$$

exist. $J : L^2(\Omega) \to L^2(\mathbb{R}^3)$ is defined by

$$(Jg)(x) := \begin{cases} j(x)g(x) & \text{for } x \in \Omega \\ 0 & \text{otherwise,} \end{cases}$$

where $j \in C_\infty(\mathbb{R}^3)$ with $j|\{x \mid |x| \leq e\} = 0$ and $j|\{x \mid |x| \geq e + 1\} = 1$. Because of local energy decay the $W^\pm$ do not depend on the special choice of j. Let $J_0 : L^2(\mathbb{R}^3) \to L^2(\Omega)$ be the adjoint of J, namely

$$(J_0 g)(x) = j(x)g(x) \quad \text{for} \quad x \in \Omega.$$

The crucial part of the existence proof for the wave operators is to show

$$(EJ_0 - J_0E_0)P_0(M) \in \mathcal{B}_1(L^2(\mathbb{R}^3), L^2(\Omega)), \tag{1.8}$$

a nuclear map from $L^2(\mathbb{R}^3)$ to $L^2(\Omega)$. P_0 is the spectral family of E_0, $M \subset \mathbb{R}$ a bounded interval, and $D := EJ_0 - J_0E_0$ is a linear first-order differential operator with C_∞ coefficients supported in $K \subset \{x|e < |x| < e + 1\}$. Equation (1.8) follows either from explicit representation of $P_0(M)U$ by means of the Fourier transform, or by exploiting the fact that, for all n, $V := P_0(M)U$ belongs to $\mathcal{D}((E_0)^n)$, and that

$$\|(E_0)^n V\| \leq c(n) \|V\|$$

holds. Using standard regularity theorems it then follows that $V \in H_5(K)$, and that

$$DP_0(M) : L^2(\mathbb{R}^3) \to \mathring{H}_4(K)$$

is a bounded operator. The inclusion map

$$i : \mathring{H}_4(K) \to L^2(K)$$

is nuclear (cf. Yosida 1974, p. 279). So

$$(EJ_0 - J_0E_0)P_0(M) : L^2(\mathbb{R}^3) \to L^2(\Omega)$$

$$(EJ_0 - J_0E_0)P_0(M) = iDP_0(M) \in \mathcal{B}_1(L^2(\mathbb{R}^3), L^2(\Omega)).$$

Therefore we obtain

$$\lim_{t\to\pm\infty} \|U(t) - U_0^{\pm}(t)\| = 0.$$

The $U_0^{\pm}(t)$ are free-space solutions with initial values $H_0^{\pm} := W^{\pm}H$. $S := W^{+}(W^{-})^{*}$ is the scattering operator. Figure 1 illustrates this. For more details compare Leis (1986, p. 112f).

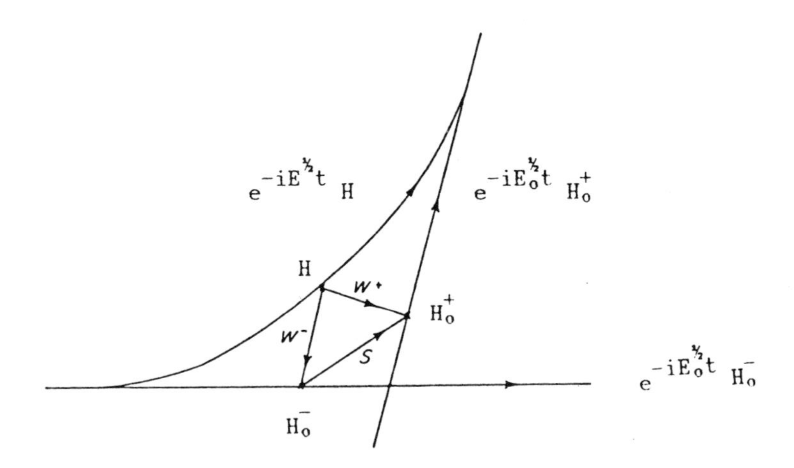

Figure 1

Finally let us remark that we could have put the elasticity equations into a system first-order in t, namely

$$\begin{pmatrix} SDU \\ U_t \end{pmatrix}_t - \begin{pmatrix} 0 & D \\ D' & 0 \end{pmatrix} \cdot \begin{pmatrix} SDU \\ U_t \end{pmatrix} = 0. \qquad (1.9)$$

We shall use this notation in the next section to be able to apply semigroup theory.

2. LINEAR THERMOELASTICITY

We use the notation of section 1. In addition to that let T be the temperature, $\Theta = T - T_0$ the temperature difference, c the specific heat, $L = (l_{ij})$ the heat conductivity tensor, and $G = (g_{ij})$ the stress-temperature tensor. The latter describes the coupling between elasticity and thermodynamics.

For the coefficients we make the usual assumptions. Let them be real-valued, bounded and measurable functions on the domain Ω, $l_{ij} = l_{ji}, g_{ij} = g_{ji}$, and

$$\exists l_1 \geq 0 \quad \forall \zeta \in \mathbb{R}^3 \quad \forall x \in \Omega \qquad \zeta_i l_{ij}(x) \zeta_j \geq l_1 |\zeta|^2,$$

$$\exists c_1 \geq 0 \quad \forall x \in \Omega \qquad c(x) \geq c_1.$$

For exterior domains in addition to this we assume the existence of positive constants $l_0, c_0 \in \mathbb{R}^+$ and $g_0 \in \mathbb{R}$ such that

$$\forall x \in \Omega_e \quad l_{ij}(x) = l_0 \delta_{ij}, \quad g_{ij}(x) = g_0 \delta_{ij}, \quad c(x) = c_0.$$

For simplicity we assume $T_0 = 1$.

The difference between linear elasticity and thermoelasticity is that we have to replace Hooke's law by the law of Duhamel-Neumann saying

$$\tau_{jk} = C_{jkmn} U_{mn} - g_{jk} \Theta, \tag{2.1}$$

and that we have to add a heat equation. Let us denote $\Gamma := (\gamma_1, \gamma_2, \ldots, \gamma_6)'$ where

$$\gamma_1 := g_{11} \quad \gamma_2 := g_{22} \quad \gamma_3 := g_{33} \quad \gamma_4 := g_{23} \quad \gamma_5 := g_{31} \quad \gamma_6 := g_{12}.$$

The linear system of thermoelasticity then reads

$$\begin{aligned} MU_{tt} - D'SDU + D'\Gamma\Theta &= 0 \\ c\Theta_t - \nabla' L \nabla \Theta + \Gamma' DU_t &= 0. \end{aligned} \tag{2.2}$$

For $\Gamma = 0$ these equations decouple into the equations of linear elasticity and the heat equation.

We are looking for solutions with finite total energy (for its definition compare the papers of Carlson (1972) and Biot (1956))

$$E(\cdot) := (DU, SDU) + (U_t, MU_t) + (\Theta, c\Theta).$$

For that reason and because we want to put the equations into a first-order system, we set

$$V := \begin{bmatrix} V_1 \\ V_2 \\ V_3 \end{bmatrix} := \begin{bmatrix} SDU \\ U_t \\ \Theta \end{bmatrix}, \quad V^0 := \begin{bmatrix} SDU^0 \\ U^1 \\ \Theta^0 \end{bmatrix}$$

$$Q := \begin{bmatrix} S^{-1} & 0 & 0 \\ 0 & M & 0 \\ 0 & 0 & c \end{bmatrix}, \quad N := \begin{bmatrix} 0 & -D & 0 \\ -D' & 0 & D'\Gamma \\ 0 & \Gamma' D & -\nabla' L \nabla \end{bmatrix},$$

and choose the Hilbert space $H(\Omega)$ to become $(L^2(\Omega))^{10}$ with scalar product

$$(V, W)_H := (V, QW)_{(L^2(\Omega))^{10}} \ .$$

We then get $E = \|V\|_H^2$. To formulate the Dirichlet problem let

$$A : \mathcal{D}(A) \subset H \to H$$
$$V \to Q^{-1} N V$$

where

$$\mathcal{D}(A) := \{V \in H \mid V_2 \in \mathring{H}_1 \wedge V_3 \in \mathring{H}_1 \wedge NV \in H\} .$$

Then we are looking for a $V \in C(R_0^+, H)$ which is a weak solution of

$$V_t + AV = 0 \text{ with } V(0) = V^0. \tag{2.3}$$

This means that for all $\Phi \in \mathring{C}(R, \mathcal{D}(A)) \cap C_1(R, \dot{H})$

$$\int_{R^+} (V, -\Phi_t + A^* \Phi)_H = (V^0, \Phi(0))_H$$

shall hold. A^* is the adjoint operator, $A^* = Q^{-1}N'$, where

$$N' := \begin{bmatrix} 0 & D & 0 \\ D' & 0 & -D'\Gamma \\ 0 & -\Gamma'D & -\nabla'L\nabla \end{bmatrix},$$

and $\mathcal{D}(A^*) = \mathcal{D}(A)$.

To derive some properties of A we start with

$$(AV, V) = (\nabla V_3, L\nabla V_3) + 2i\ \mathrm{Im}\{(V_1, DV_2) + (\Gamma'DV_2, V_3)\}$$

yielding

$$\mathrm{Re}(AV, V) = (\nabla V_3, L\nabla V_3) \geqq l_1|V_3|_1^2 \geqq 0.$$

Furthermore

$$\mathcal{N}(A) = \mathcal{D}_0' \times \mathcal{O} \times \mathcal{O}$$

where $\mathcal{D}_0' := \{U \in (L^2(\Omega))^6 \mid D'U = 0\}$ and $\mathcal{N}(A^*) = \mathcal{N}(A)$. A is a closed operator and

$$\mathcal{H} = \overline{\mathcal{R}(A)} \oplus \mathcal{N}(A^*) = \overline{\mathcal{R}(A)} \oplus \mathcal{N}(A)$$

$$\overline{\mathcal{R}(A)} = \overline{SD\mathring{H}_1} \oplus (L^2)^3 \oplus L^2.$$

$\overline{\mathcal{R}(A)}$ and $\mathcal{N}(A)$ reduce A. Thus it is possible to restrict ourselves to $\overline{\mathcal{R}(A)}$ when dealing with (2.3).

Let $\lambda \in \mathbb{C}$ with $\mathrm{Re}\,\lambda < 0$ and $(A-\lambda)V = 0$. Then we get

$$(\nabla V_3, L\nabla V_3) - \mathrm{Re}\,\lambda\,\|V\|^2 = 0$$

and thus $V = 0$. By the same argument $\mathcal{N}(A^* - \bar{\lambda}) = 0$. Therefore $(A-\lambda)^{-1}$ exists and $\mathcal{H} = \overline{\mathcal{R}(A-\lambda)}$. Let $(A-\lambda)V = F$. Then again

$$(\nabla V_3, L\nabla V_3) - \text{Re}\,\lambda \,\|V\|^2 \leq |(V, F)|$$

or

$$\|(A-\lambda)^{-1}\| \leq -1/\text{Re}\,\lambda.$$

Thus we have proved

$$\mathbb{C}^- := \{\lambda \in \mathbb{C} \mid \text{Re}\,\lambda < 0\} \subset \rho(A),$$

and from semigroup theory we get the existence of a family $H(t)$, $t \geq 0$, of bounded linear operators such that:

1. $H(0) = \text{id}$, $H(s+t) = H(s)H(t)$;
2. $\|H(t)\| \leq 1$;
3. $H(t)A \subset AH(t)$;
4. $\forall V \in \mathcal{H}$, $t \to H(t)V$ is continuous;
5. $\forall V \in \mathcal{D}(A)$, $t \to H(t)V$ is differentiable;
6. $\forall V^0 \in \mathcal{D}(A)$, $V(t) := H(t)V^0 \in C(\mathbb{R}_0^+, \mathcal{D}(A)) \cap C_1(\mathbb{R}_0^+, \mathcal{H})$ is the unique solution of $V_t + AV = 0$ with $V(0) = V^0$.

Thus we have solved the Dirichlet problem. Other initial-boundary value problems can be treated similarly. Some immediate conclusions are possible

$$\forall t \;\|V(t)\| \leq \|V^0\|$$

is obvious. $\|\nabla V_3(t)\| \to 0$ as $t \to \infty$ also holds. Let

$$\mathcal{I} := \{V \in \mathcal{D}(A) \mid \forall t \geq 0 \;\; \|H(t)V\| = \|H^*(t)V\| = \|V\|\}$$

where $H^*(t)$ is the semigroup generated by A^*. It is the adjoint of $H(t)$. $\mathcal{I}$ is a closed subspace of $\mathcal{D}(A)$ with respect to $\|\cdot\|_A := (\|\cdot\|^2 + \|A\cdot\|^2)^{\frac{1}{2}}$ so that

$$\mathcal{D}(A) = \mathcal{I} \oplus \mathcal{I}^{\perp}. \tag{2.4}$$

I and $I^\perp$ are invariant under $H(t)$ and $H^*(t)$. Let $V^0 \in \mathcal{D}(A)$, $V(t) := H(t)V^0$ and

$$V^0 = V_1^0 + V_2^0$$

according to (2.4). Then

(i) $\forall t \geqq 0 \; \|H(t)V_1^0\| = \|H^*(t)V_1^0\| = \|V_1^0\|$

(ii) $\forall r > 0 \;\; \lim_{t\to\infty} \|V(t) - H(t)V_1^0\|_{\Omega_r} = 0.$

For details compare Racke (1987). These results describe the asymptotic behaviour of $V(t)$ for bounded domains Ω. The problem remains to characterize I. It can be shown that I is the $\|\cdot\|_A$-closure of the span of eigenfields belonging to purely imaginary eigenvalues. These eigenvalues give rise to undamped vibrations. It is interesting to note that they exist iff

$$M^{-1}D'SDU + \lambda U = 0 \text{ with } \Gamma'DU = 0$$

has nontrivial solutions $U \in \mathring{H}_1$. An example for the existence of such eigenfunctions has been given where Ω is the unit circle (in $\mathbb{R}^2$).

In the following let Ω be an exterior domain. In that case from (ii) we get local energy decay

$$\forall r > 0 \; \lim_{t\to\infty} \|V(t)\|_{\Omega_r} = 0.$$

To obtain further results we first discuss the corresponding free-space problem assuming an isotropic medium. Thus let M = id, L = id, $c = 1$, $g_{ij} = \gamma\delta_{ij}$, $\gamma \in \mathbb{R}$, and

$$D'SD = -\mu \text{ rot rot} + \nu \text{ grad div}, \quad D'\Gamma = \gamma \text{ grad}$$

where μ,ν are positive constants. Furthermore let $H_0 := (L^2(\mathbb{R}^3))^{10}$ and A_0 be defined in analogy to A. We already know

$$H_0 = R(A_0) \oplus N(A_0) = (SDH_1 \times (L^2)^3 \times L^2) \oplus (\mathcal{D}_0' \times 0 \times 0).$$

Both subspaces reduce A_0. To get a further reduction we define

$$H_0^s := \overline{SD(H_1 \cap \mathcal{D}_0)} \times \mathcal{D}_0 \times 0$$

$$H_0^p := \overline{SD(H_1 \cap R_0)} \times R_0 \times L^2$$

where

$$(L^2(\mathbb{R}^3))^3 = \mathcal{D}_0 \oplus R_0$$

$$R_0 = \{U \in (L^2(\mathbb{R}^3))^3 \mid \text{rot } U = 0\} = \overline{\nabla H_1}$$

$$\mathcal{D}_0 = \{U \in (L^2(\mathbb{R}^3))^3 \mid \text{div } U = 0\}.$$

We then obtain a decomposition into solenoidal and potential fields, precisely

$$H_0 = H_0^s \oplus H_0^p \oplus N(A_0), \tag{2.5}$$

all subspaces reducing A_0. Therefore our equations (2.2) decompose into

$$U_{tt}^s + \mu \text{ rot rot } U^s = 0 \tag{2.6}$$

and

$$\begin{aligned} U_{tt}^p - \nu \text{ grad div } U^p + \gamma \text{ grad } \Theta &= 0 \\ \Theta_t - \Delta\Theta + \gamma \text{ div } U_t^p &= 0. \end{aligned} \tag{2.7}$$

Equations (2.6) are the Maxwell equations. Their solutions are undamped vibrations and may be discussed in analogy to section 1. For details compare Leis (1986, p. 146f). The coupling constant γ only appears in (2.7), which are the more interesting equations.

To discuss the spectrum of A_0 and the asymptotic behaviour of the solutions of $V_t + A_0V = 0$ one can use the Fourier transform, and after straightforward but lengthy calculations obtains detailed results. Let $\Lambda_1 := i\mathbb{R}$ and

$\Lambda_2 := \mathrm{R}_0^+$. Then we get for the spectrum Λ of A_0

$$\Lambda = \Lambda^s \cup \Lambda^p(\gamma) \cup \{0\}$$

where $\lambda = 0$ is the point eigenvalue originating from $N(A_0)$. $\Lambda^s = \Lambda_1$ due to (2.6). For Λ^p we obtain

$$\Lambda^p(\gamma) := \{\lambda \in \mathbb{C} \mid \exists q \in \mathrm{R}^3 \quad \Delta(\lambda,q,\nu,\gamma) = 0\}$$

$$= \Lambda_2 \cup \Lambda_3(\gamma)$$

where $\Lambda_3(0) = \Lambda_1$ and

$$\Delta(\lambda,q,\nu,\gamma) := -\lambda^3 + |q|^2\lambda^2 - (\nu + \gamma^2)|q|^2\lambda + \nu|q|^4.$$

The characteristic polynomial of $\hat{A}_0$, the Fourier transform of A_0, reads

$$0 = \hat{A}_0^3 \cdot (A_0^2 + \mu\, |q|^2)^2 \cdot \Delta(\hat{A}_0,q,\nu,\gamma). \tag{2.8}$$

It is interesting to give the analogous polynomials for the corresponding problems in R^2 and R^1. They read

$$0 = \hat{A}_0 \cdot (\hat{A}_0^2 + \mu\, |q|^2) \cdot \Delta(\hat{A}_0,q,\nu,\gamma) \quad \text{in } \mathrm{R}^2 \tag{2.9}$$

$$0 = \Delta(\hat{A}_0,q,\nu,\gamma) \qquad\qquad \text{in } \mathrm{R}^1. \tag{2.10}$$

Thus in R^1 we neither get a null space nor a vibrating component.

Figure 2 shows some $\Lambda_3(\gamma)$ for $\nu = 1$ in the upper half-plane. The curves are symmetric with respect to the x-axis. For more details compare Leis (1980, 1981, 1986, p. 231f). We also remark that Racke (1986) introduced generalized eigenfunctions and that he gave an eigenfunction expansion.

Knowing the spectrum of A_0 we can also calculate the asymptotic behaviour of the solutions of $V_t + A_0V = 0$. Let

$$V^0 = V^{0s} + V^{0p} + V^{00}$$

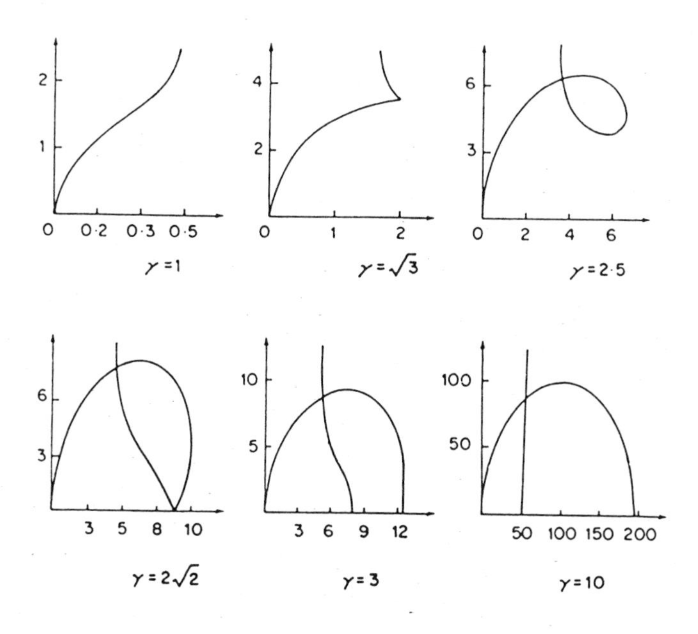

Figure 2

according to (2.5). Then

$$V(t) = H(t)V^{0s} + H(t)V^{0p} + V^{00}$$

where

(i) $V^{00} = H(t)V^{00}$ is stationary;

(ii) $H(t)V^{0s}$ is an undamped vibration of Maxwell type - the asymptotic behaviour is similar to that given in section 1;

(iii) let $\gamma = 0$ - then the first two components of $H(t)V^{0p}$ are undamped vibrations also and the third component is a solution of the heat equation and thus vanishes as $t \to \infty$;

(iv) let $\gamma \neq 0$ - then $\lim_{t\to\infty} \|H(t)V^{0p}\| = 0$.

The asymptotic behaviour of the solutions of only a few boundary value problems has been given so far. Assuming an isotropic medium these are boundary conditions which are compatible with the decomposition into solenoidal and potential fields. In that case we obtain results similar to those we got for the free-space problem. Such boundary conditions are

$$(n \times U)|_{\partial\Omega} = 0 \wedge (\mathrm{div}\ U)|_{\partial\Omega} = 0 \wedge \Theta|_{\partial\Omega} = 0$$

or

$$(n \cdot U)|_{\partial\Omega} = 0 \quad \wedge \ (n \times \mathrm{rot}\ U)|_{\partial\Omega} = 0 \wedge \partial\Theta/\partial n|_{\partial\Omega} = 0,$$

where U is the elastic displacement vector again.

3. SOME NONLINEAR PROBLEMS

In this last section I want to indicate some results on the existence of solutions of nonlinear problems global in t for sufficiently small and smooth data. Let us start with the free-space problem of elasticity (homogeneous isotropic medium in R^3), and write the linear system as a first-order system, cf. (1.9); symbolically

$$V_t + AV = 0 \text{ with } V(0) = V^0. \tag{3.1}$$

Let the nonlinear problem be given by

$$V_t + AV = F(V,\nabla V) \text{ with } V(0) = V^0, \tag{3.2}$$

and let P be the projector on $N(A)$. Then naturally we assume $PV^0 = 0$ and $PF(V,\nabla V) = 0$ for all V, ∇V.

We consider (3.2) to be a perturbation of (3.1) and assume

$$|F(V,\nabla V)| \leqq c(|V| + |\nabla V|)^3 \tag{3.3}$$

for $|V| + |\nabla V|$ small. Then Klainerman (1982) and Klainerman and Ponce (1983) proved that for sufficiently small and smooth V^0 a solution of (3.2) global in t exists, and that asymptotically it behaves like a solution of the linear system (3.1). In this case, therefore, wave operators can be introduced again (cf. (3.14)).

The idea of the proof is to combine the usual local existence theorem with an *a priori* estimate of the solution. One starts by deriving sharp $L^p - L^q$ estimates concerning the asymptotic behaviour of solutions of the linear equation (3.1). Let

$$V(t) := e^{-At} V^0$$

be the solution of (3.1). Then

$$\|V(t)\|_{L^2} = \|V^0\|_{L^2}$$

immediately follows. From explicit knowledge of the fundamental solution of (3.1) one also obtains

$$\|V(t)\|_{L^\infty} \leq c_\infty (1 + t)^{-1} \|V^0\|_{L^1,3} \tag{3.4}$$

(derivatives up to the third order have to be taken on the right-hand side). The general $L^p - L^q$ estimate then is a consequence of the usual interpolation inequalities between these extreme cases. It reads

$$\|V(t)\|_{L^q} \leq c_q (1 + t)^{-1+2/q} \|V^0\|_{L^p, N_p} \tag{3.5}$$

where $q \geq 2$, $1/p + 1/q = 1$ and $3(2-p)/p \leq N_p \leq 3$.

Next a local existence theorem is used to obtain a solution $V \in C([0,T],H_s) \cap C_1([0,T],H_{s-1})$ of the nonlinear equation (3.2) in an interval $[0,T]$, $T > 1$. Then

$$\forall t \in [0,T] \quad \|V(t)\|_{L^\infty} + \|\nabla V(t)\|_{L^\infty} < 1$$

follows, and $\|V^0\|_s$ is assumed to be small ($\|\cdot\|_s$ denotes the L^2-norm of all derivatives up to the order s, $s \geq 3$). To prove the local existence theorem one first estimates the solutions of the linearized problem, and their derivatives, and then applies the contraction principle. Details can be found in Klainerman (1980, p. 94f).

In the third step "high-energy estimates" are derived by elementary but tricky partial integration and applying Gronwall's lemma. This means that one can estimate high derivatives of the local solution (∂ represents any derivative, $\partial_t, \partial_1, \partial_2$, or ∂_3)

$$\|V(t)\|_s \leq c_s \|V^0\|_s \exp(c_s \int_0^t [\|V(r)\|^2_{L^\infty} + \|\partial V(r)\|^2_{L^\infty}]dr). \tag{3.6}$$

The fourth step is essential. Defining with $\tau := 7 + N_{6/5}$, $\rho := \tau + N_{6/5}$, and $\sigma := 2 + \rho$

$$M_\tau(T) := \sup_{t\in[0,T]} (1 + t)^{2/3} \|V(t)\|_{L^6_{,\tau}}$$

one shows that a constant M_0, not depending on T, exists such that

$$M_\tau(T) \leq M_0 \tag{3.7}$$

holds for small $V^0 \in L^2_{,\sigma} \cap L^{6/5}_{,\rho}$. To derive this estimate one starts from

$$V(t) = e^{-At}V^0 + \int_0^t e^{-A(t-r)}F(V,\nabla V)(r)dr. \tag{3.8}$$

Using the calculus inequality (cf. Klainerman 1980, 1982)

$$\|F(V,\nabla V)\|_{L^{6/5}_{,\rho}} \leq c\|\nabla V\|_{L^2_{,\sigma}} \cdot \|V\|^2_{L^6_{,\tau}} \tag{3.9}$$

and

$$\|V\|_{L^\infty} + \|\partial V\|_{L^\infty} \leq c\,\|V\|_{L^6_{,\tau}}, \tag{3.10}$$

which follows from Sobolev's inequality and the differential equation, one obtains from (3.8) and (3.5) with $q = 6$, $p = 6/5$, and (3.6)

$$x \leq c\delta(1 + x^2e^{cx^2}) \tag{3.11}$$

where $x := M_\tau(T)$ and $\|V^0\|_{L^2_{,\sigma}} + \|V^0\|_{L^{6/5}_{,\rho}} \leq \delta$ is small. This yields (3.7).

Estimate (3.7) then leads to the *a priori* estimate for the local solution $V(t)$, we wanted to prove, namely

$$\exists\ K_\sigma \text{ independent of } T\ \forall t \in [0,T]\ \|V(t)\|_\sigma \leq K_\sigma \|V^0\|_\sigma, \tag{3.12}$$

assuming $\|V^0\|_{L^2_{,\sigma}} + \|V^0\|_{L^{6/5}_{,\rho}}$ to be sufficiently small.

Thus we can reapply the local existence theorem to obtain the desired global solution $V \in C([0,\infty),H_\sigma) \cap C_1([0,\infty),H_{\sigma-1})$ for

$$\|V^0\|_{L^2_{,\sigma}} + \|V^0\|_{L^{6/5}_{,\rho}} \leq \delta$$

sufficiently small. From (3.7) and (3.10)

$$\|V(t)\|_{L^6_{,\tau}} + \|V(t)\|_{L^\infty} + \|\partial V(t)\|_{L^\infty} = O(t^{-2/3}) \tag{3.13}$$

as $t \to \infty$ follows. Defining

$$V^+(t) := V(t) + \int_t^\infty e^{-A(t-r)}F(V,\nabla V)(r)dr,$$

V^+ is a solution of the linear equation $V_t + AV = 0$, and the previous results immediately lead to

$$\lim_{t\to\infty} \|V(t) - V^+(t)\|_{L^\infty} = \lim_{t\to\infty} \|V(t) - V^+(t)\|_{L^6_{,\tau}} = 0. \tag{3.14}$$

It should be remarked that similar statements hold in $\mathbb{R}^n$, $n \geq 2$. When $n \geq 6$ quadratic terms on the right-hand side of (3.3) are allowed. In the case of the wave equation $n \geq 4$ suffices to obtain that result. This has been shown by Klainerman (1985) who improved the estimate (3.4) by replacing the L^1-norm on the right-hand side by L^2-norms of ΓV^0. Γ consists of differential operators which leave solutions of the wave equation invariant (cf. also Christodoulou 1986, John 1987). In $\mathbb{R}^3$, and with quadratic behaviour in (3.3), global smooth solutions of the wave equation may exist when a "null condition" is fulfilled (Klainerman 1986); an example can be found in Klainerman (1980, p. 45). John (1981) has shown, however, that generally they blow. Plane waves and radial solutions also blow. In $\mathbb{R}^1$ solutions always blow (John 1974, 1976).

The next problem is to treat initial-boundary value problems. So far this has only been done by Shibata and Tsutsumi (1986) for the wave equation with constant coefficients and "nontrapping" domains with Dirichlet boundary condition. Here also the asymptotic behaviour of the solutions of the linear equation has to be given first, and therefore the "nontrapping" condition is used.

Let us now switch to thermoelasticity. We again write the underlying equations (cf. (2.2)-(2.3)) in the form

$$V_t + AV = F(W) \text{ with } V(0) = V^0. \tag{3.15}$$

W is a vector composed of V and certain derivatives of V; $F := (0,f_2,f_3)'$, where $f_2 = f_2(BV_1,\nabla BV_1,V_3,\nabla V_3)$ and $f_3 = f_3(BV_1,\nabla BV_1,\nabla V_2,V_3,\nabla V_3,\nabla^2 V_3)$. The medium again is assumed to be homogeneous and isotropic, and

$$B : \overline{SDL^2_{,1}} \to L^2, \text{ bounded}$$

$$U \to \nabla\cdot(SD)^{-1}U.$$

Let us first deal with the corresponding problem in $\mathbb{R}^1$. From (2.10) we know that in the linear case we do not get vibrations. The coupling term appears isolated so that a study in $\mathbb{R}^1$ should help to understand the general situation in $\mathbb{R}^3$. Since there are no vibrations in $\mathbb{R}^1$ we expect that heat dissipation is strong enough to prevent a solution from blowing up at least for small and smooth data. This was proved by Zheng and Shen (1987). On the other hand Dafermos and Hsiao (1986) gave an example with blow-up for large and smooth data.

For bounded domains in $\mathbb{R}^1$ Slemrod (1981) showed the existence of global solutions for small and smooth data and special boundary conditions, namely $U_x|_{\partial\Omega} = 0$ and $\Theta|_{\partial\Omega} = 0$ or $U|_{\partial\Omega} = 0$ and $\Theta_x|_{\partial\Omega} = 0$ (here again U denotes the elastic displacement vector). The Dirichlet initial-boundary value problem was treated by Racke (1988a,b). He is able to show local existence for bounded and exterior domains. High-energy estimates, however, are still missing, and hence the question whether global solutions exist is still open.

An exterior initial-boundary value problem in $\mathbb{R}^1$, again $U_x|_{\partial\Omega} = 0$ and

$\Theta|_{\partial\Omega} = 0$, has been treated by Jiang (1988). He uses the Fourier sine and cosine transform to obtain $L^p - L^q$ estimates for the linear problem. He also proves high-energy estimates, and he is able to show the existence of solutions global in t for small and smooth data. Jiang is also capable of treating the boundary condition $U|_{\partial\Omega} = 0$ and $\Theta_x|_{\partial\Omega} = 0$.

In $\mathbb{R}^3$ the situation is more difficult since we anticipate vibrations (cf. (2.8)-(2.10)). On the other hand we already know that global solutions exist for the hyperbolic part. So we again expect the existence of solutions global in t. Local existence follows from a result of Kawashima (1983), whereas Racke (1988c) showed the existence of global solutions for small and smooth data essentially assuming

$$F(W) = F_1(W) + F_2(BV_1,V_3)$$

where

$$F_1(W) = O(|W|^3) \text{ and } F_2(BV_1,V_3) = O(\|BV_1\|^3 + \|V_3\|^2)$$

for small $|W|$. The term F_2 only appears in the heat equation. To do so he also derived $L^p - L^q$ estimates for the linear equation and gave high-energy estimates. Initial-boundary value problems have not been dealt with as yet.

Let me end up by indicating some open problems. So far we have assumed a homogeneous and isotropic medium. Little is known for inhomogeneous or anisotropic media. In any case it would be most important to prove global existence and possible uniqueness of weak solutions for large data.

In $\mathbb{R}^1$, as we know, the parabolic component dominates. Systems of conversation laws in $\mathbb{R}^1$ have been studied and solved using Glimm's difference scheme (Glimm 1965; cf. Smoller 1983). It should be looked at whether this method works in thermoelasticity also.

In $\mathbb{R}^3$, however, we meet with a parabolic and a hyperbolic component. The existence of certain weak solutions for parabolic equations has been proved (cf. von Wahl 1985). Solutions of hyperbolic equations with large data, however, generally develop singularities at a time $t = T_0$, the life span, which was proved by John (1981); cf. Majda (1984). T_0 depends on the largeness of the initial data; $T_0 = \infty$ for small and smooth data and cubic nonlinearity, as we have seen before. Generally the existence of global weak solutions has not yet been shown, and so far one does not have a clear idea of what happens in nonlinear thermoelasticity.

REFERENCES

Belopolskii, A.L. and M.Sh. Birman (1968). The existence of wave operators in the theory of scattering with a pair of spaces. Math. USSR Izv. 2, 1117-1130.

Biot, M.A. (1956). Thermoelasticity and irreversible thermodynamics. J. Appl. Phys. 27, 240-253.

Carlson, D.E. (1972). Linear Thermoelasticity. Handbuch der Physik, VIa/2, 297-346, Springer-Verlag, Berlin.

Christodoulou, D. (1986). Global solutions of nonlinear hyperbolic equations for small data. Commun. Pure Appl. Math. 39, 267-287.

Dafermos, C.M. (1968). On the existence and the asymptotic stability of solutions to the equations of linear thermoelasticity. Arch. Rat. Mech. Anal. 29, 241-271.

Dafermos, C.M. and L. Hsiao (1986). Development of singularities in solutions of the equations of nonlinear thermoelasticity. Q. Appl. Math. 44, 463-474.

Glimm, J. (1965). Solutions in the large for nonlinear hyperbolic systems of equations. Commun. Pure Appl. Math. 18, 95-105.

Jiang, S. (1988). Global existence and asymptotic behaviour of smooth solutions in one-dimensional nonlinear thermoelasticity. Thesis, University of Bonn. Bonner Mathematische Soleriften No. 192.

John, F. (1974). Formation of singularities in one-dimensional nonlinear wave propagation. Commun. Pure Appl. Math. 27, 377-405.

John, F. (1976). Delayed singularity formation in solutions of nonlinear wave equations in higher dimensions. Commun. Pure Appl. Math. 29, 649-681.

John, F. (1981). Blow-up for quasi-linear wave equations in three space dimensions. Commun. Pure Appl. Math. 34, 29-51.

John, F. (1987). Existence for large times of strict solutions of nonlinear wave equations in three space dimensions for small initial data. Commun. Pure Appl. Math. 40, 79-109.

Kato, T. (1976). Perturbation Theory for Linear Operators. Springer-Verlag, Berlin.

Kawashima, S. (1983). Systems of a hyperbolic-parabolic composite type, with applications to the equations of magnetohydrodynamics. Thesis, Kyoto University.

Klainerman, S. (1980). Global existence for nonlinear wave equations. Commun. Pure Appl. Math. 33 43-101.

Klainerman, S. (1982). Long-time behavior of solutions to nonlinear evolution equations. Arch. Rat. Mech. Anal. 78, 73-98.

Klainerman, S. (1985). Uniform decay estimates and the Lorentz invariance of the classical wave equation. Commun. Pure Appl. Math. 38, 321-332.

Klainerman, S. (1986). The null condition and global existence to nonlinear wave equations. In Nonlinear Systems of Partial Differential Equations in Applied Mathematics. Lectures in Appl. Math. 23, 293-326. American Math. Soc.

Klainerman, S. and G. Ponce. (1983). Global, small amplitude solutions to nonlinear evolution equations. Commun. Pure Appl. Math. 36, 133-141.

Kupradze, V.D. (1979). Three Dimensional Problems in Elasticity and Thermoelasticity. North-Holland, Amsterdam.

Leis, R. (1970). Zur Theorie elastischer Schwingungen in inhomogenen Medien. Arch. Rat. Mech. Anal. 39, 158-168.

Leis, R. (1980). Aussenraumaufgaben in der linearen Elastizitätstheorie. Math. Meth. Appl. Sci. 2, 379-396.

Leis, R. (1981). Über das asymptotische Verhalten thermoelastischer Wellen im $\mathbb{R}^3$. Math. Meth. Appl. Sci. 3, 312-317.

Leis, R. (1986). Initial Boundary Value Problems in Mathematical Physics. B.G. Teubner, Stuttgart, and John Wiley, Chichester.

Majda, A. (1984). Compressible Fluid Flow and Systems of Conservation Laws in Several Space Variables. Springer-Verlag, New York.

Pearson, D.B. (1978). A generalization of the Birman trace theorem. J. Funct. Anal. 28, 182-186.

Racke, R. (1986). Eigenfunction expansions in thermoelasticity. J. Math. Anal. Appl. 120, 596-609.

Racke, R. (1987). On the time-asymptotic behaviour of solutions in thermoelasticity. Proc. R. Soc. Edinburgh 107A, 289-298.

Racke, R. (1988a). Initial boundary value problems in one-dimensional nonlinear thermoelasticity. Math. Meth. Appl. Sci. 10, 517-529.

Racke, R. (1988b). Initial boundary value problems in thermoelasticity. In: Partial Differential Equations and Calculus of Variations, Lecture Notes in Mathematics No. 1357. Springer-Verlag, Berlin, 341-358.

Racke, R. (1988c). On the Cauchy problem in nonlinear 3 d-thermoelasticity. SFB 256 Preprint No. 26, Bonn.

Rellich, F. (1943). Über das asymptotische Verhalten der Lösungen von $\Delta u + \lambda u = 0$ in unendlichen Gebieten. Jber. Dt. Math.-Verein. 53, 57-65.

Shatah, J. (1982). Global existence of small solutions to nonlinear evolution equations. J. Diff. Equ. 46, 409-425.

Shibata, Y. and Y. Tsutsumi. (1986). On a global existence theorem of small amplitude solutions for nonlinear wave equations in an exterior domain. Math. Z. 191, 165-199.

Slemrod, M. (1981). Global existence, uniqueness, and asymptotic stability of classical smooth solutions in one-dimensional nonlinear thermoelasticity. Arch. Rat. Mech. Anal. 76, 93-134.

Smoller, J. (1983). Shock waves and reaction-diffusion equations. Springer-Verlag, New York.

Sommerfeld, A. (1949). Vorlesungen über theoretische Physik, Bd. II. Akad. Verlagsgesellschaft Geest & Portig, Leipzig.

von Wahl, W. (1985). The Equation of Navier-Stokes and Abstract Parabolic Equations. Vieweg, Braunschweig.

Weck, N. (1969). Aussenraumaufgaben in der Theorie stationärer Schwingungen inhomogener elastischer Körper. Math. Z. 111, 387-398.

Yosida, K. (1974). Functional Analysis. Springer-Verlag, Berlin.

Zheng, S. and W. Shen. (1987). Global solutions to the Cauchy problem of a class of quasi-linear hyperbolic parabolic coupled systems. Sci. Sinica. To appear.

R. Leis
Institut für Angewandte Mathematik
der Universität Bonn
Wegelerstrasse 10,
D-5300 Bonn 1,
Federal Republic of Germany

H.A. LEVINE

The long-time behaviour of solutions of reaction-diffusion equations in unbounded domains: a survey

1. INTRODUCTION

Let $D \subset R^N$ be a domain with a piecewise smooth boundary, or else $D = R^N$. There has been, it is fair to say, more than a passing interest in positive solutions of the problem:

$$
\text{(P)}\qquad
\begin{aligned}
&u_t = \Delta u + u^p \quad \text{in } D \times (0,T), \quad (p > 1)\\
&u(x,t) = 0 \qquad (x,t) \in \partial D \times (0,T)\\
&u(x,0) = u_0(x) \quad x \in D.
\end{aligned}
$$

It is known that when D is a bounded domain, not all solutions are global, a result due to Kaplan [11] and, as a consequence of more general considerations, to Levine [13], who showed that if

$$E(u_0) \equiv \frac{1}{2}\int_D |\nabla u_0|^2\, dx - \frac{1}{p+1}\int_D u_0^{p+1} dx < 0, \tag{1.1}$$

then the solutions (P) cannot be global.† (D need not be bounded.) More recently several authors have examined the precise nature of how the solution fails to be global. Beginning with Ball [2], several authors [4,5,8,19,20,21], to cite just a few, have studied the pointwise blow-up of solutions of (P) for large initial data.

In the study of pointwise blow-up, these authors restricted p to satisfy

* This research was supported by the Air Force Office of Scientific Research under Grant No. AFOSR 88-0031. The United States Government is authorized to reproduce and distribute reprints for governmental purposes not withstanding any copyright notation therein.

† That is, the solution cannot remain in $H_0^1(D) \cap L^{p+1}(D)$ for all time.

$$p < \begin{cases} (N+2)/(N-2) & \text{if } N > 2 \\ \infty & \text{if } N = 1,2, \end{cases} \tag{1.2}$$

whereas the (weaker) global nonexistence results of [13] do not require this restriction. (In some cases $p = (N + 2)/(N - 2)$ is included in the study of single point blow-up [8].)

On the other hand, for any $p > 1$, when D is bounded, (P) can have nontrivial global solutions for any initial values u_0 for which $\Delta u_0 + u_0^p \leqq 0$. These global solutions will decay to zero if the set

$$S = \{f \in H_0^1(D) \mid \Delta f + f^p = 0 \text{ in } D, \quad f = 0 \text{ on } \partial D, \ f \geqq 0\}$$

is either $\{0\}$ or if

$$u_0(x) \leqq \inf\{f(x) \mid f \in S\} \quad x \in D$$

with strict inequality on an open subset of D. If (1.2) holds, it can be shown that S is not trivial. See [18] for example.

Fujita [6] was perhaps the first to examine (P) on all of R^N. He proved the following interesting result.

<u>THEOREM 1</u> [6]: (a) If $1 < p < 1 + 2/N$, then (P) does not possess nontrivial global solutions. (b) If $p > 1 + 2/N$, then there are global positive solutions of (P).

Case (a) is often called the blow-up case while (b) is called the global existence case. (In [1,10,12] and later in [20], it was established that $p = 1 + 2/N$ belonged to the blow-up case.) Such a result may be called a "Fujita-type" blow-up theorem. It is the purpose of this talk to discuss other such Fujita-type results for parabolic equations.

It should perhaps be remarked that John, Glassey and others [26-34] have obtained partial "Fujita-type" results for hyperbolic problems such as

$$
(H) \quad \begin{array}{ll} u_{tt} = \Delta u + |u|^p & (x,t) \in R^N \times (0,T) \\ u(x,0) = u_0(x) & x \in R^N \\ u_t(x,0) = v_0(x) & x \in R^N, \end{array}
$$

although the results are not as complete as they are for (P) (as is to be expected). For example, in [31] it was shown that if $1 < p < p_0(N)$ where $P_0(N)$ is the larger root of $(N - 1)p^2 - (N + 1)p - 2 = 0$ then (H) has no nontrivial global solutions. On the other hand, Glassey [26,27] (when $N = 2$) and John [28] (when $N = 3$) have shown that for $p > p_0(N)$ small data, nontrivial, global solutions exist. Schaeffer [30] showed that both $p_0(2)$ and $p_0(3)$ belong to the blow-up case. More recently, in [32,33] it is shown that if $N = 2,3,4,\ldots$

$$
p > p_1(N) := (N^2 + 3N - 2)/N(N - 1),
$$

then small data, nontrivial global solutions of (H) do exist. Thus, to our knowledge, there remains a gap for (H) when $N > 3$.†

We turn now to a discussion of (P) in other unbounded domains.

2. THE FIRST RESULTS OF MEIER

In [16], Meier considered (P) when, for fixed $k \in [1,N]$, k an integer,

$$
D_k = \{x \mid x_1 > 0,\ldots,x_k > 0\}.
$$

The results of Meier are somewhat more general than we present here in that he considers the equation

$$
u_t = \Delta u + t^q u^p \text{ in } D_k \times [0,\infty)
$$

where $q \geq 0$, $p > 1$. However, we shall state them here only in the case of $q = 0$ for purposes of comparison with the results of Fujita et al. for (P). Let $p(k,N) = 1 + 2/(N + k)$.

THEOREM 2 [16]: (a) If $1 < p < p(k,N)$, (P) has no nontrivial global solutions

---

† I am told that Sideras has closed this gap but the work is unpublished.

(b) If $p > p(k,N)$ there are global, bounded solutions of (P) which decay uniformly to zero on D_k as $t \to +\infty$.

It is not known whether or not $p(k,N)$ falls into the blow-up case for $k > 0$.[1]

The Fujita-type results of Fujita, Meier, Weissler and others depend heavily on specific properties of the Green's function for the heat equation in R^N or D_k. For example, if $H(x,t)$ denotes the Green's function, then

$$H(0,t) = (2\pi t)^{-N/2} \quad (D = R_N)$$

and

$$\|H(\cdot,t)\|_{L^1(R^N)} = 1 \ (D = R_N),$$

properties which played an important role in the arguments of [16] and [20]. It is the second property, which fails if $D = D_k$, that prevented Meier from extending Weissler's argument to this case.[1]

Because, for most geometries, the Green's function is not readily found, it is desirable to have alternate methods available for investigating asymptotic properties of solutions. For example, the argument of Kaplan [11] for proving blow-up (global nonexistence) depended on the positivity of the first Dirichlet eigenfunction for Δ on bounded domains. Such an argument, which works also for hyperbolic problems on bounded domains, fails on unbounded domains.

Recently, however, Bandle and Levine [3] have modified Kaplan's argument to obtain blow-up results of Fujita type for other unbounded domains. They have also obtained Fujita-type global existence results for "large" p. More recently Levine and Meier [14] have improved upon some of these large p results. The arguments of [3, 14] for large p also avoid the use of the Green's function for the heat equation. We recently learned of some related results of Kavian and others which are obtained by different methods. We turn next to a discussion of these results.

---

(1) But see Section 6 and the note added in proof.

3. THE RESULTS OF BANDLE AND LEVINE

We consider the case for which D is a cone in $\mathbb{R}^N$ with vertex at the origin (for convenience). That is, we write, for $x \in \mathbb{R}^N$, $x = (r,\underset{\sim}{\theta})$ where $r \in (0,\infty)$ and $\underset{\sim}{\theta} \in \Omega$ where $\Omega \subset S^{N-1}$ is a submanifold of the unit sphere with boundary, $\partial\Omega$, smooth enough to permit integration by parts. We assume also that $\partial\Omega$ has positive (N - 2)-dimensional measure.

Let ω_1 be the smallest Dirichlet eigenvalue for the Laplace-Beltrami operator on Ω. Let $\gamma_\pm$ denote the positive and negative roots respectively of

$$\gamma^2 + \gamma(N - 2) - \omega_1 = 0.$$

Explicitly,

$$\gamma_\pm = (1/2)\ \{2 - N \pm [(N - 2)^2 + 4\omega_1]^{1/2}\}.$$

The following result is given in [3] as Theorem 2.3 and Theorem 7.5.

THEOREM 3 [3]: (a) If

$$1 < p < 1 + 2/(2 - \gamma_-) = \underline{p} \tag{3.1}$$

then no nontrivial, nonnegative, almost regular solution of (P) can be global in time. (b) Let

$$\bar{p} := \min(1 + 2/N,\ 1 + 2/(-\gamma_-)). \tag{3.2}$$

If

$$p > \bar{p} \tag{3.3}$$

then (P) has nontrivial global, almost regular solutions.

REMARK: In [3], a generalization of (a) in which u^p is replaced by $f(u)$ is also given. See also [35].

A solution of (P) is almost regular on $Q_T = D \times [0,T)$ if

(i) $u \in C^2(Q_T) \cap C^0(\bar{Q}_T - D \times \{T\})$,

(ii) for all $k > 0$ and $t \in [0,T)$, $\liminf_{r\to\infty} e^{-kr} \int_\Omega (|u| + |u_r|)dS_{\underset{\sim}{\theta}} = 0$,

(iii) there is a sequence $\{r_n\}_{n=1}^{\infty}$, $r_n \to 0$ such that

$$\lim_{n\to\infty} \int_\Omega [r_n^{N-1}|u_r(r_n,\underset{\sim}{\theta},t)| + r_n^{N-2}|u(r_n,\underset{\sim}{\theta},t)|]dS_{\underset{\sim}{\theta}}.$$

Let us write out the conditions (3.1), (3.3). We see that we always have $\underline{p} < 1 + 2/N$ unless $\omega_1 = 0$ in which case $\underline{p} = 1 + 2/N$.

Moreover, $\bar{p} = 1 + 2/N$ if and only if $\omega_1 \geq 2N$, i.e. if and only if Ω is "small". Thus, for fixed $p > 1$, in accordance with what is known for bounded domains, small cones are more stable than large ones. In the case $N = 2$ and $\Omega = (0,\gamma\pi)$ we have

$$\underline{p} = 1 + 2\gamma/(1 + 2\gamma)$$

$$\bar{p} = \min(2, 1 + 2\gamma),$$

so that if $\gamma < 1/2$, the statement (b) is an improvement in the range of p for which we have global solutions over what we would obtain if we apply the Fujita result.

The idea behind the proof of (a) is the following: Since we cannot apply the argument of Kaplan directly, we let $\psi(\underset{\sim}{\theta})$ be the first Dirichlet eigenfunction of the Laplace-Beltrami operator, $\Delta_{\underset{\sim}{\theta}}$, with $\psi > 0$ on Ω and

$$\int_\Omega \psi(\underset{\sim}{\theta})dS_{\underset{\sim}{\theta}} = 1.$$

We then let, for $m,k > 0$,

$$\phi(r,\underset{\sim}{\theta}) = C^{-1}r^m e^{-kr}\psi(\underset{\sim}{\theta}),$$

where $C = k^{-(m+N)}\Gamma(m + N)$ so that

$$\int_D \phi \, dx = 1.$$

If

$$(k^2 + \lambda)[m^2 + (N - 2)m - \omega_1] \geqq [m + (1/2)(N - 1)]^2 k^2 \tag{3.4}$$

then it is easy to show that

$$\Delta\phi + \lambda\phi \geqq 0$$

and that

$$F(t) = \int_D u\phi \, dx$$

satisfies

$$F'(t) \geqq -\lambda F(t) + (F(t))^p \tag{3.5}$$

in view of the almost regularity of u. Consequently u will not be global in time if

$$F(0) > \lambda^{1/(p-1)}. \tag{3.6}$$

Now (3.4), (3.5) and (3.6) will simultaneously hold provided

$$m^2 + (N - 2)m - \omega_1 > 0 \tag{3.7}$$

$$\beta \equiv \frac{\lambda}{k^2} = \frac{m + \omega_1 + (1/4)(N - 1)^2}{m^2 + (N - 2)m - \omega_1} \tag{3.8}$$

and

$$k^{-[2/(p-1)-(m+N)]} \int_0^\infty r^{m+N-1} e^{-kr} \left(\int_\Omega \psi(\underset{\sim}{\theta}) u_0(r,\underset{\sim}{\theta}) dS_{\underset{\sim}{\theta}} \right) dr > \Gamma(m+N)\beta^{1/(p-1)}. \tag{3.9}$$

Thus, if

$$2 - N - \gamma_- = \gamma_+ < m < 2/(p - 1) - N$$

and β is fixed, we can choose k (and hence λ) so small that (3.7), (3.8) and (3.9) hold.

Concerning (b), if $\bar{p} = 1 + 2/N$, the result follows from the Fujita result by comparison.

If $\bar{p} = 1 + 2/(-\gamma_-)$, the argument is more subtle and depends upon the following sequence of lemmas. (If $p \geqq (N + 1)/(N - 3)$, the result again follows from the Fujita result by comparison since $1 + 2/N < (N + 1)/(N - 3)$.) Thus, it suffices to consider only the case $1 + 2/(-\gamma_-) < p < (N + 1)/(N - 3)$. The lemmas are of interest in their own right. (They are Theorems 3.2, 7.4, 6.1, 4.6 of [3].)

THEOREM 4 [3]: If

$$1 - 2/\gamma_- < p < \begin{cases} (N+1)/(N-3) & N > 3 \\ \infty & N = 2,3, \end{cases}$$

then there is a singular stationary solution of the form

$$u_s(r,\underset{\sim}{\theta}) = r^{-2/(p-1)}\alpha(\underset{\sim}{\theta}) \tag{3.10}$$

where $\alpha(\underset{\sim}{\theta}) > 0$ in Ω and solves

$$\Delta_{\underset{\sim}{\theta}}\alpha + \nu\alpha + \alpha^p = 0 \text{ in } \Omega$$

$$\alpha = 0 \text{ on } \partial\Omega,$$

and $\nu = 2/(p - 1)[2/(p - 1) + 2 - N]$. If $1 < p \leqq \bar{p}$, there is no singular solution of the form given in (3.10). When $N = 2$, there is at most one solution of this form.

A regular stationary solution of (P) is a function $w \geqq 0$, $w \in C^2(D) \cap C^0(\bar{D})$ such that

$$\text{(S)} \qquad \begin{aligned} \Delta w + w^p &= 0 \quad \text{in } D \\ w &= 0 \quad \text{on } \partial D. \end{aligned}$$

REMARK: The only stationary solutions $w(x)$ of (P) such that

$$\lambda^{2/(p-1)} w(\lambda x) = w(x)$$

for all $\lambda > 0$, are of the form (3.10).

Then, the remainder of (b) follows from

LEMMA 5 [3]: If p satisfies the conditions of Theorem 4 and if

$$0 \leqq u_0 \leqq \min\{r^{\varepsilon}, u_s\} \tag{3.11}$$

for some $\varepsilon > 0$ and some solution u_s of the form, then the solution of (P) is global in time.

If p is further restricted, more is true.

THEOREM 6 [3]: If

$$1 - 2/\gamma_- < p < \begin{cases} (N+2)/(N-2) & N > 2 \\ \infty & N = 2, \end{cases}$$

and if (3.11) holds, then not only is u global but also

$$\lim_{t\to\infty} u(x,t) = 0, \quad x \in D.$$

To prove Lemma 5, one defines

$$\bar{u}(r, \underset{\sim}{\theta}) = \inf\{u_s, \psi\}$$

where $\psi = Ar^m \alpha(\underset{\sim}{\theta})$ and $m > 0$ satisfies

$$m^2 + (N - 2)m - \nu < 0,$$

where $\nu(> \omega_1)$ is given in Theorem 4. The function $\bar{u}$ is a supersolution of (S) in the weak sense. The result then follows by comparison principles.

The proof of Theorem 6 is more difficult because it depends on the following lemma:

LEMMA 7 [3]: Let p satisfy the hypothesis of Theorem 6. Then there are no nonnegative solutions of (S), w(x), such that $w \leqq u_s$ in D.

The main idea of the proof of Lemma 7 is to consider the one-parameter family

$$w(x;\lambda) = \lambda^{2/(p-1)} w(\lambda r, \underset{\sim}{\theta}).$$

One can show (but not easily) that if this family has an envelope on any subcone, then w is singular at r = 0. If it fails to have an envelope on any subcone it follows that

$$G(r,\underset{\sim}{\theta}) \equiv w_r + \frac{2}{p-1}\frac{w}{r} \geqq 0$$

on the cone with strict inequality somewhere. If we then set

$$\ell(r,\underset{\sim}{\theta}) = r^{2/(p-1)} w(r,\underset{\sim}{\theta})$$

and

$$L(r) = \int_\Omega \ell(r,\underset{\sim}{\theta})\alpha(\underset{\sim}{\theta})dS_{\underset{\sim}{\theta}},$$

a calculation leads to

$$L_{rr} + (N - 1 - 2q)L_r \geqq 0 \quad (q \equiv 2/(p-1)).$$

Whence, for $r > r_1$, where $L_r(r_1) > 0$

$$L(r) \geqq L(r_1) + \text{const}\begin{cases} r^{2q+1-N} & p < (N+2)/(N-2) \\ \ln r & p = (N+2)/(N-2), \end{cases}$$

where the constant is positive. From this, we see that L(r) is unbounded on (r_1,∞). On the other hand, since $w < u_s$,

$$L(r) \leq \int_{\Omega} \alpha^2(\underset{\sim}{\theta})dS_{\underset{\sim}{\theta}}$$

and obvious contradiction.

In [3], we also prove

THEOREM 8 [3]: If $1 < p < 1 - 2/\gamma_-$ then there are no regular stationary solutions of (S) except $w \equiv 0$.

4. RELATED RESULTS

The technique of [3] can also be applied to study (P) in domains which are exterior to a bounded region. Let $D^c = R^N - D$ be a bounded region. Then in [3] we prove the following. (Theorems 8.2, 8.3, of [3].)

THEOREM 9 [3]: (a) If $1 < p < 1 + 2/N$, (P) has no nontrivial nonnegative global solutions. (b) If $p > 1 + 2/N$ there are nontrivial bounded global solutions. If $N \geq 3$, $p > 1 + 2/(N - 2)$, $0 \in D^c$ and

$$u(x,0) \equiv u_0(x) \leq \lambda|x|^{-2/(p-1)}$$

where

$$\lambda \equiv \left[\frac{2}{p - 1}\left(N - 2 - \frac{2}{p - 1}\right)\right]^{1/(p-1)}$$

then u is a global solution of (P). If $N/(N-2) < p \leq (N + 2)/(N - 2)$, $u(x,t)$ decays to zero pointwise at $t \to \infty$.

Extensions of this result to other problems (including those with convection-like terms) are given in [36]. Theorem 9(a,b) settles an old conjecture of Fujita.

The proof of the nonexistence result is similar to the proof of the result when D is a cone. The first part of (b) follows from Fujita's result and comparison. The second part of (b) follows by arguments similar to those used to prove Theorem 4 and Lemma 5.

In [9], *stationary* solutions of the problem

$$(P_\sigma)\quad \begin{cases} u_t = \Delta u + |x|^\sigma u^p & \text{in } D \times [0,T) \\ u = 0 & \text{on } \partial D \times (0,T) \\ u(x,0) = u_0(x) & \text{on } D \end{cases}$$

(where $u_0 \geqq 0$) were considered when $D = R^N$. In [3], we established the following when D is a cone and $\sigma > -2$.

<u>THEOREM 10</u> [3]: Let

$$\underline{p} = 1 + (2 + \sigma)/(2 - \gamma_-)$$

$$\bar{p} = 1 + (2 + \sigma)/(-\gamma_-).$$

(a) If $1 < p < \underline{p}$, there are no nontrivial global regular solutions of (P_σ).
(b) If

$$\bar{p} < p < \begin{cases} (N+1)/(N-3) & N \geqq 4 \\ \infty & N = 2,3, \end{cases}$$

there are singular stationary solutions of (P_σ) of the form

$$u_s = r^{-(2+\sigma)/(p-1)}\alpha(\underset{\sim}{\theta}).$$

If

$$\bar{p} < p < \min\left\{\frac{N+1}{N-3}, \frac{N+2+2\sigma}{N-2}, \infty\right\}$$

then regular solutions of (P_σ) with $u(x,0) \leqq \min(r^\varepsilon, u_s)$ for some $\varepsilon > 0$ are global and decay to zero pointwise as $t \to \infty$. If

$$\bar{p} < \frac{N+2+2\sigma}{N-2} < p < \begin{cases} (N+1)/(N-3) & N \geqq 4 \\ \infty & N = 3 \end{cases}$$

then regular solutions with $u(x,0) \leqq \min(r^\varepsilon, u_s)$ for some $\varepsilon > 0$ are global.

(c) If $1 < p < \bar{p}$, (P_σ) has no stationary solutions except $u \equiv 0$ and no singular solutions of the above form.

The proofs of the statements in Theorem 10 are very similar to the proofs of the corresponding statements when $\sigma = 0$ except in the case of the first statement. In most of these arguments, the quantity $2/(p - 1)$ is replaced by $(2 + \sigma)/(p - 1)$.

For the statement (a), we let

$$\delta = \sigma/(p - 1)$$

and put

$$u = |x|^{-\delta} v.$$

Then the differential equation becomes

$$\begin{aligned} v_t = {} & r^{-(N-2\delta-1)} \frac{\partial}{\partial r}\left(r^{N-2\delta-1} \frac{\partial v}{\partial r}\right) \\ & + r^{-2}[\Delta_{\underset{\sim}{\theta}} v - \delta(N - 2 - \delta)v] + v^p. \end{aligned} \tag{4.1}$$

If we set

$$F(t) = \int_0^\infty \int_\Omega v\phi r^{N-1} \, dS_{\underset{\sim}{\theta}} dr$$

we find that (3.5) holds provided (3.4) holds with m replaced by $m + \delta$. The condition that both (3.5) and (3.6) hold for all λ sufficiently small becomes

$$\gamma_+ - \delta < m < 2/(p - 1) - N$$

and (a) follows.

5. THE RESULTS OF LEVINE AND MEIER

Bandle and Levine [3] observed that when $D = D_k$,

$$\underline{p} = p(\dot{k},N),$$

so that, in view of Meier's early result [15], $\underline{p}$ is the cutoff between the blow-up case and the global existence case. Meier then conjectured that this was true for *every* cone, i.e. that $\underline{p} = \bar{p}$.

In [17] he showed that, when $N = 2$ and $\gamma = 1/n$, $n = 1,2,\ldots,\underline{p}$ is the cutoff between (a) and (b). He also showed there that if $N = 3$ and

$$\Omega = \{(\phi,\theta) \in S^2 \mid 0 < \phi < \pi/n, \quad 0 < \theta < \pi\}$$

or

$$\Omega = \{(\phi,\theta) \in S^2 \mid 0 < \phi < \pi/n, \quad 0 < \theta < \pi/2\},$$

there $\underline{p}$ is again the cutoff for (a), (b). The values of $\underline{p}$ for these cases are given in Table 1.

Meier's original arguments were modified in [17] but they also utilized the Green's function which can be constructed for such domains by the method of images.

He also showed there that if D is any domain, there is a critical exponent p^*, $p^* \geqq 1$, such that if $p > p^*$ there are always nontrivial global solutions of (P) while if $1 < p < p^*$ (when $p^* > 1$) there are no nontrivial global solutions. (When D is bounded, $p^* = 1$.)

Careful inspection of Meier's arguments led him and Levine to conclude that it should be possible to prove that for cones, $p^* = \underline{p}$ in all cases. This they were able to do by showing that if $\nu := \gamma_+ + \frac{1}{2}(N - 2)$ and $t_0 > 0$

$$w(r,\underset{\sim}{\theta},t) := r^{-(N-2)/2} \int_0^\infty e^{-\lambda(t+t_0)} J_\nu(\sqrt{\lambda}r) J_\nu(\sqrt{\lambda}) d\lambda \psi(\underset{\sim}{\theta}), \tag{5.1}$$

Table 1. Summary of Fujita-type results for (P)

	Domain	p	$\bar{p}$	Notes and References	
1.	R^N	$1 + 2/N$	$1 + 2/N$	(1)	[1]
2.	$R^N - D^c$ (D^c bounded)	$1 + 2/N$	$1 + 2/N$	(2,3)	[3]
3.	$D = (0,\infty) \times \Omega$				
	$\Omega \subset S^{N-1}$, $\omega_1 > 0$	$1 + 2/(2-\gamma_-)$	$\begin{cases} 1 + 2/(-\gamma_-) & \omega_1 \geqq 2N \\ 1 + 2/N & \omega_1 \leqq 2N \end{cases}$	(5,6)	[3]
			$1 + 2/(2-\gamma_-)$	(4)	[14]
	Examples of above				
4.	$N = 2$,				
	$D = (0,\infty) \times (0,\gamma\pi)$	$1 + 2\gamma/(1+2\gamma)$	$\begin{cases} 1 + 2\gamma & \gamma < \frac{1}{2} \\ 2 & \gamma \geqq \frac{1}{2} \end{cases}$	(2)	[3]
4.	$D = (0,\infty) \times (0,\gamma\pi)$				
	$\gamma = 1/n$, $n = 1,2,\ldots$	$1 + 2/(n+2)$	$1 + 2/(n+2)$	(2)	[17]
6.	$N = 3$, $D = (0,\infty) \times \Omega$				
	$\Omega = \{(\phi,\theta) \mid 0 < \phi < \pi/n, 0 < \theta < \pi\}$	$1 + 2/(n+3)$	$1 + 2/(n+3)$	(2)	[17]
7.	$N = 3$, $D = (0,\infty) \times \Omega$				
	$\Omega = \{(\phi,\theta) \mid 0 < \phi < \pi/n, 0 < \theta < \pi/2\}$	$1 + 2/(n+4)$	$1 + 2/(n+4)$	(2)	[17]

	Domain	$\underline{p}$	$\bar{p}$	Notes and References
8.	$D = \{x \mid x_1 \geq 0, \ldots, x_k \geq 0\}$ $1 \leq k \leq N$	$1 + 2/(N+k)$	$1 + 2/(N+k)$	(2) [15]
9.	D bounded	1	1	(7)

Notes.

(1) $\underline{p}$ belongs to the blow up case.

(2) $\underline{p}$ belongs to the blow up case, [22].

(3) If $N > 2$, and $p > N/(N-2)$ there is a singular stationary solution in $\mathbf{R}^N - \{0\}$. Solutions starting under this stationary solution are global.

(4) ω_1 is the first Dirichlet eigenvalue of the Laplace-Beltrami operator. If periodic boundary conditions are imposed in the angular variables ($\Omega = S^{N-1}$) then $\omega_1 = 0$.

(5) Singular stationary solutions exist if

$$1 - 2/\gamma - < p < \begin{cases} (N+1)/(N-3) & N > 3 \\ \infty & N = 2,3 \end{cases}$$

Solutions for which $u_0 \leq \min(r^{\varepsilon}, u_s)$ for some $\varepsilon > 0$ are global in time.

(6) Meier has conjectured that $\underline{p}$ belongs to the blow up case and is the boundary of the blow up case.

(7) $\underline{p} = 1$ belongs to the no blow up case.

then w is a positive* solution of the heat equation vanishing on the boundary of the cone. Then one can find a function $\beta(t)$ defined on $[0,\infty)$ such that

$$\bar{u}(r,\underset{\sim}{\theta},t) = \beta(t)w(r,\underset{\sim}{\theta},t) \tag{5.2}$$

is a global, positive, supersolution of (P). See [14] for details.

Meier has also conjectured that $\underline{p}$ belongs to the blow-up case but this problem remains open(1). (See section 6 below, however.)

The gap between $\underline{p}$ and $\bar{p}$ can be closed for (P_σ) as well, at least for $\sigma \geqq 0$. We have

THEOREM 11 [14]: Let $\underline{p}$ be as in Theorem 10 and $\sigma \geqq 0$.

(a) If $1 < p < \underline{p}$, $(P\ )$ has no nontrivial global solutions.

(b) If $p > \underline{p}$, then (P_σ) has nontrivial global solutions.

The extra condition arises through the construction of a supersolution for (P_σ). One has to require that

$$\limsup_{t\to+\infty} (t + t_0)^{\frac{(N+\gamma_+)}{(2+\sigma)}} \left[\sup_{r>0,\underset{\sim}{\theta}\in\Omega} r^\sigma w^{p-1}(r,\underset{\sim}{\theta},t) \right] < \infty.$$

This leads to the condition that $\sigma \geqq 0$.

We obtain the following corollary (Corollaries 3.2, 3.4 of [14].)

COROLLARY 12 [14]: If u solves (P_σ) where $p > \underline{p}(\ )$, $\sigma \geqq 0$ and if $u(r,\underset{\sim}{\theta},t) \leqq \beta(t)w(r,\underset{\sim}{\theta},t)$ for some t_0, then for all $r > 0$,

$$\limsup_{t\to+\infty} t^{(N+\gamma_+)/(2+\sigma)} \sup_{\theta\in\Omega} u(r,\underset{\sim}{\theta},t) \leqq cr^{-\sigma/(p-1)}$$

for some constant c depending only upon $\beta(0)$, t_0 and geometry.

* The integral can be evaluated. It has the value

$$(t + t_0)^{-1}\exp[-(1+r^2)/4(t+t_0)]I_\nu(r/2(t+t_0))$$

where I_ν is the modified Bessel function of order ν. See Watson, A Treatise on the Theory of Bessel Functions, Cambridge University Press (1922), p. 395.

When $\sigma = 0$, we may replace $\sup_\Omega$ by $\sup_D$. In particular, the L^∞ norm of the solution decays *faster* than $(t + t_0)^{-1/(p-1)-\varepsilon}$ for some $\varepsilon > 0$.

We remark finally that our existence and nonexistence results take place in the space

$$\{f \mid f = 0 \text{ on } \partial D, \int_D e^{-k|x|}(|f(x)| + |\nabla f(x)|)dx < \infty \quad \text{for all } k > 0\}.$$

This is a larger space than $H_0^1(D)$ as our singular solution, $r^{-2/(p-1)}\alpha(\underset{\sim}{\theta})$, belongs to this space if $p > 1 + 2/(N - 1)$ but not to $H_0^1(D)$. (The inclusion follows from Schwarz's inequality.)

6. THE RESULTS OF ESCOBEDO AND KAVIAN

In [22,23,24], the authors examine (P) from the point of view of L^2 theory. They introduce the weighted Hilbert spaces

$$L^2(D,K) = \{f \mid \int_D |f|^2 K(y)dy < \infty\},$$

$$H_0^1(D,K) = \{f \mid f, \nabla f \in L^2(D,K), f = 0 \text{ on } \partial D\},$$

where $K(y) = \exp(\frac{1}{4}|y|^2)$. On this space they consider the initial-boundary value problem for

$$v_s + Lv = |v|^{p-1}v + (p - 1)^{-1}v$$

where

$$v(s,y) := e^{s/(p-1)}u(e^s - 1, e^{s/2}y), \tag{6.1}$$

u solves (P) and where

$$Lf \equiv -\Delta f - \frac{1}{2}y\cdot\nabla f = K^{-1}\nabla\cdot(K\nabla f)$$

denotes the self-adjoint operator which results from the change of variables (6.1). If λ_1 is the smallest eigenvalue of L on $H_0^1(D,K)$, then [22] $\lambda_1 \geqq \frac{1}{2}N$, $N/2$ being the smallest eigenvalue of L on $\mathbb{R}^N$.

They treat the problem from the point of view of potential well theory [25]. We quote their principal results for our cones below. (It is assumed that $u(\cdot,t) \in H_0^1(D)$ unless otherwise specified.) The following result is Theorem 4.5 of [22].

THEOREM 13 [22]: Let $1 < p < (N + 2)/(N - 2)$ and D be a *convex* cone.

(i) If $p \leq 1 + 1/\lambda_1$ and $u(\cdot,0) \geq 0$, then $u(\cdot,t)$ blows up in finite time.

(ii) If $p \geq 1 + 1/\lambda_1$ and $u(0) \in H_0^1(D,K)$

$$E(0,K) :\equiv \frac{1}{2}\int_D |\nabla u(0)|^2 K(y)dy - \frac{1}{(p+1)} \int_D |u(0)|^{p+1} K(y)dy$$

$$- \frac{1}{2(p-1)} \int_\Omega |u(0)|^2 K(y)dy \leq 0, \tag{6.2}$$

then $u(t)$ blows up in finite time.

(iii) If $p > 1 + 1/\lambda_1$ there exists $u_0 \in H_0^1(D,K)$ $u_0 \geq 0$, $u_0 \not\equiv 0$, such that $u(t)$ is global in time.

(iv) If $p > 1 + 1/\lambda_1$ and $u(t)$ is a global solution, then

$$\limsup_{t\to+\infty} t^{1/(p-1)} \|u(t)\|_{C^2(D)} < \infty.$$

We see from this that when our cones are convex we have the p belongs to the blow-up case.[1] (For all cones, one can show directly that $\lambda_1 = \frac{1}{2}(N + \gamma_+)$.)

Also, as we see from Corollary 12, some solutions do decay more rapidly than $t^{-1/(p-1)}$. Are there any solutions for which

$$\lim_{t\to\infty} t^{1/(p-1)} \|u(t)\|_{L^\infty}$$

is a finite, nonzero number, i.e. solutions which decay exactly like $t^{-1/(p-1)}$ in L^∞?

We see also that the range of p is somewhat restricted here since $p < (N + 2)/(N - 2)$ and that the class of solutions for which blow-up occurs is smaller than that considered in [3].

The singular solutions we have constructed here do not belong to the Hilbert space when D is a cone. Moreover, they exist over a range of p which neither includes nor is contained in $(1, (N + 2)/(N - 2))$ when $N > 2$. Thus some additional structure is lost when the Hilbert space approach is taken.

The result (ii) holds for any $p > 1$ when (6.2) holds. This is a consequence of [13] since $f(v) = |v|^{p-1}v + v(p - 1)$ satisfies the structure conditions of [13] and since L is positive definite on $H_0^1(D,K)$.

Finally, (iv) does not improve Theorem 6 because of the extra condition that $u(\cdot) \in H_0^1(D,K)$.

REFERENCES

I. Parabolic Problems

[1] D.J. Aronson and H.F. Weinberger, Multidimensional nonlinear diffusion arising in population genetics, Adv. Math. 30 (1978), 33-76.

[2] J.M. Ball, Remarks on blow up and nonexistence theorems for nonlinear evolutionary equations, Q. J. Math. Oxford Ser. 28 (1977), 473-486.

[3] C. Bandle and H.A. Levine, On the existence and nonexistence of global solutions of reaction-diffusion equations in sectorial domains, Trans. Amer. Math. Soc. (in print).

[4] P. Baras and L. Cohen, Complete blow up for the solution of a semilinear heat equation, J. Funct. Anal. 71 (1987), 142-174.

[5] A. Friedman and B. McLeod, Blow up of positive solutions of semilinear heat equations, Indiana Univ. Math. J. 34 (1985), 425-447.

[6] H. Fujita, On the blowing up of solutions of the Cauchy problem for $u_t = u + u^{1+\alpha}$, J. Fac. Sci., Tokyo Sect. IA, Math. 16 (1966), 105-113.

[7] H. Fujita, On some nonexistence and nonuniqueness theorems for nonlinear parabolic equations, Proc. Symp. Pure Math. 18 (1969), 105-113.

[8] B. Gidas and J. Spruck, Global and local behavior of positive solutions of nonlinear elliptic equations, Commun. Pure Appl. Math. 23 (1981), 525-598.

[9] Y. Giga and R.V. Kohn, Asymptotically self-similar blow up of semilinear heat equations, Commun. Pure Appl. Math. 38 (1985), 297-319.

[10] K. Hayakawa, On nonexistence of global solutions of some semilinear parabolic equations, Proc. Japan Acad. 49 (1973), 503-525.

[11] S. Kaplan, On the growth of solutions of quasilinear parabolic equations, Commun. Pure Appl. Math. 16 (1963), 305-333.

[12] K. Kobayashi, T. Siaro, and H. Tanaka, On the blowing up problem for semilinear heat equations, J. Math. Soc. Japan 29 (1977), 407-424.

[13] H.A. Levine, Some nonexistence and instability theorems for solutions of formally parabolic equations of the form $Pu_t = -Au + F(u)$, Arch. Rat. Mech. Anal. 51 (1973), 371-386.

[14] H.A. Levine and P. Meier, The value of the critical exponent for reaction-diffusion equations in cones Arch. Rat. Mech. Anal. (in print).

[15] P. Meier, Existence et non-existence de solutions globales d'une équation de la chaleur semi-linéaire: extension d'un théorème de Fujita, C.R. Acad. Sci. Paris Ser. I 303 (1986), 635-637.

[16] P. Meier, Blow up of solutions of semilinear parabolic differential equations, ZAMP 39 (1988), 135-149.

[17] P. Meier, On the critical exponent for reaction-diffusion equations Arch. Rat. Mech. Anal. (in print).

[18] D.H. Sattinger, Topics in Stability and Bifurcation Theory, in Lect. Notes Math., Vol. 309, Springer, New York, 1973.

[19] D.H. Sattinger, Monotone methods in nonlinear elliptic and parabolic boundary value problems, Indiana Univ. Math. J. 21 (1972), 979-1000.

[20] F.B. Weissler, Existence and nonexistence of global solutions for a semilinear heat equation, Israel J. Math. 38 (1981), 29-40.

[21] F.B. Weissler, An L blow-up estimate for a nonlinear heat equation, Commun. Pure Appl. Math 38 (1985), 291-295.

[22] O. Kavian, Remarks on the large time behavior of a nonlinear diffusion equation, Ann. Inst. Henri Poincaré, Analyse Nonlineaire 4 (1987), 423-452.

[23] M. Escobedo and O. Kavian, Asymptotic behavior of positive solutions of a nonlinear heat equation (preprint).

[24] M. Escobedo and O. Kavian, Variational problems related to self similar solutions of the heat equation, J. Nonlinear Anal. TMA (in press).

[25] L.E. Payne and D.H. Sattinger, Saddle points and instability of non-linear hyperbolic equations, Israel J. Math. 22 (1975), 273-303.

II. Hyperbolic Problems

[26] R. Glassey, Finite time blow up for solutions of nonlinear wave equations, Math. Z. 177 (1981), 323-340.

[27] R. Glassey, Existence in the large for u = F(u) in two space dimensions, Math. Z. 178 (1981), 233-261.

[28] F. John, Blow up of solutions of nonlinear wave equations in three dimensions, Man. Math. 28 (1979), 235-268.

[29] J. Schaeffer, Finite time blow up for $u_{tt} = \Delta u = H(u_r, u_t)$ in two space dimensions. Commun. PDE 11 (5) (1986), 513-545.

[30] J. Schaeffer, The equation $u_{tt} - \Delta u = |u|^p$ for the critical value p, Proc. R. Soc. Edinburgh 101 A (1985), 31-44.

[31] T. Sideras, Nonexistence of global solutions of semilinear wave equations in high dimensions, J. Diff. Eqns. 52 (1984), 378-406.

[32] Ta-Tsien Li and Yun Mei Chen, Solutions regulieres global du probleme de Cauchy pour les équations des ondes non linéaires, C.R. Acad. Sci. Paris Sér. I Math. 305 (1987), 171-174.

[33] Ta-Tsien Li and Yun-Mei Chen, Initial value problems for nonlinear wave equations, Commun. PDE 13 (1988), 383-422.

[34] M. Rammaha, Nonlinear wave equations in high dimensions, Proc. Int. Conf. on Theory and Applications of Differential Equations, Ohio University, Athens, Ohio, March 21-25, 1988 (to appear).

III. Additional References

[35] C. Bandle, Blow up in exterior domains, Proc. Nancy Conf. on Nonlinear Evolution Equations 1988 (to appear).

[36] C. Bandle and H.A. Levine, Fujita type results for convective reaction diffusion equations in exterior domains (to appear).

[37] H.A. Levine and P. Meier, A blow up result for the critical exponent in Cones (to appear).

H.A. Levine
Department of Mathematics
Iowa State University
Ames, Iowa 50011,
U.S.A.

Added in Proof: After this paper had been written, Meier and the author were able to establish that for any cone, the number p given in Theorem 3 belongs to the blow up case. The argument is a modification of that of Weissler [20]. By careful scrutiny of Weissler's arguments coupled with judicious use of inequalities we were able to avoid the difficulties to which we alluded in the discussion following Theorem 2. The result will appear in [37].

J. MAWHIN

Bifurcation from infinity and nonlinear boundary value problems

1. INTRODUCTION

Let H be a real Hilbert space and $L : D(L) \subset H \to H$ a linear self-adjoint operator with compact resolvent. If $\sigma(L)$ denotes the spectrum of L (a pure point spectrum $\{\lambda_i\}_{i\in J}$ with no finite accumulation point), the following result concerning the solvability of the equation

$$Lu - \lambda u = h \tag{1}$$

for $\lambda \in \mathbb{R}$ and $h \in H$ is well-known:

(i) If $\lambda \notin \sigma(L)$, problem (1) has a unique solution for each $h \in H$.

(ii) If $\lambda = \lambda_i \in \sigma(L)$, problem (1) has a solution if and only if $h \in N_i^\perp$, where $N_i = N(L - \lambda_i I)$.

Moreover, we can describe as follows the set of solutions (λ,u) of (1) in the neighbourhood of $\lambda_i \in \sigma(L)$. As $H = N_i \oplus N_i^\perp$, we can write each element u of H in the form $u = \bar{u} + \tilde{u}$, with $\bar{u} \in N_i$ and $\tilde{u} \in N_i^\perp$, and (1) is equivalent to the system

$$(L - \lambda_i I)u - (\lambda - \lambda_i)\tilde{u} = \tilde{h},$$

$$- (\lambda - \lambda_i)\bar{u} = \bar{h}.$$

Letting

$$L_i = (L - \lambda_i I)|_{D(L) \cap N_i^\perp} : D(L) \cap N_i^\perp \to N_i^\perp$$

(a bijection), we can write, for $0 < |\lambda - \lambda_i|$ sufficiently small, the unique solution $u(\lambda)$ of (1) in the form (with I_i the identity on $N_i^\perp$)

$$u(\lambda) = (\lambda_i - \lambda)^{-1}\bar{h} + (I_i - (\lambda - \lambda_i)L_i^{-1})^{-1}L_i^{-1}\tilde{h},$$

and the second term of the right-hand member has a limit when $\lambda \to \lambda_i$. Therefore, if $\bar{h} \neq 0$, i.e. if $H \notin N_i^{\perp}$,

$$\|u(\lambda)\| = \{|\lambda_i-\lambda|^{-2} \|\bar{h}\|^2 + \|[I_i - (\lambda-\lambda_i)L_i^{-1}]^{-1}L_i^{-1}\tilde{h}\|^2\}^{1/2} \to +\infty$$

if $\lambda \to \lambda_i$. On the other hand, if $\bar{h} = 0$, i.e. if $h \in H_i^{\perp}$, then

$$u(\lambda) \to L_i^{-1}\tilde{h} \in N_i^{\perp}$$

if $\lambda \to \lambda_i$, and the set of solutions of (1) for $\lambda = \lambda_i$ is given by $L_i^{-1}\tilde{h} + N_i$. Thus, the bifurcation diagram $(\lambda, \|u\|)$ for the solutions of (1) has, for λ close to λ_i, the shape indicated on figure 1.

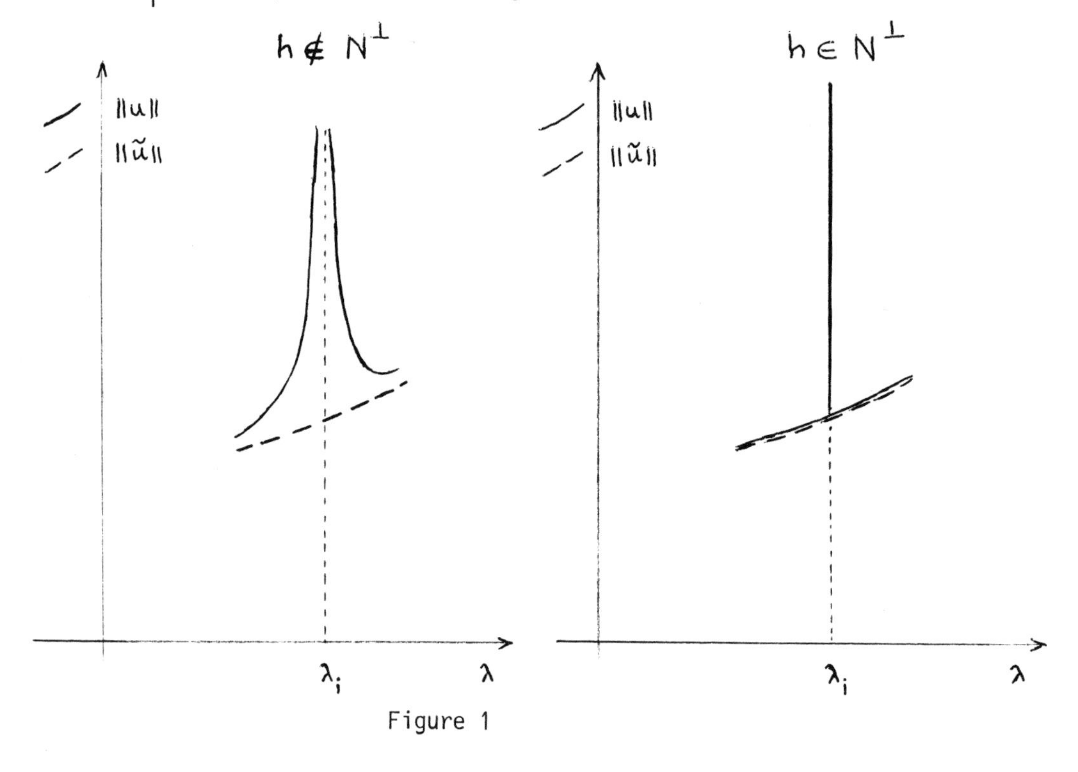

Figure 1

It may be of interest to study cases where h is replaced by a nonlinear operator N on H and the closest situation to the inhomogeneous problem (1) is that where N is continuous and bounded on H, i.e. $\|Nu\| \leq C$ for all $u \in H$ and some $C > 0$. The corresponding nonlinear equation

$$Lu - \lambda u = Nu \qquad (2)$$

is equivalent, when $\lambda \notin \sigma(L)$, to the fixed point problem

$$u = (L - \lambda I)^{-1}Nu,$$

whose right-hand member is a completely continuous operator on H with bounded range. The Schauder fixed point theorem immediately implies the existence of at least one solution. For the case where $\lambda = \lambda_i \in \sigma(L)$, i.e. for the equation

$$Lu - \lambda_i u = Nu, \qquad (3)$$

the existence question is more delicate (as for the linear case), but conditions have been introduced in the late 1960s by Lazer and his coworkers [5,6,7] which may be viewed as some type of nonlinear version of the condition $h \in N_i^{\perp}$ in the linear case. The reader may consult [2] or [8] for more recent references to those conditions generally referred as Landesman-Lazer conditions.

Now, when λ_i has an odd multiplicity, a result of Krasnosel'skii [3,4], subsequently refined and extended by Stuart [13], Toland [14], Rabinowitz [12] and others, implies for equations (2) with the above assumptions the existence of a continuous branch of solutions going off from (λ_i,∞) in the sense that for each $\varepsilon > 0$, there is a ball B centred at zero in H such that, on the boundary Γ of each bounded open neighbourhood of B, there is a solution (λ,u) of (2) with $\lambda \in (\lambda_i - \varepsilon, \lambda_i + \varepsilon)$ and $u \in \Gamma$. Applied to the linear problem (1), such a result does not distinguish the situations where $h \in N_i^{\perp}$ or $h \notin N_i^{\perp}$. A striking difference between the two situations seems to be the existence, when $h \in N_i$, of an *a priori* bound independent of λ for the solutions u of (1) when $\lambda \in [\lambda_i - \delta, \lambda_i + \delta]\setminus\{\lambda_i\}$, for some $\delta > 0$. Following the spirit of some recent joint work with Schmitt [10,11], we shall describe in this paper the use of bifurcation from infinity and of the existence of *a priori* estimates near λ_i for the solutions of (2) in the obtention of existence and multiplicity results in the nonlinear case.

2. NONLINEAR EIGENVALUE PROBLEMS WITH SUBLINEAR NONLINEARITY

Let X be a real Banach space, $L : D(L) \subset X \to X$ a Fredholm linear operator with index zero, and $N : X \to X$ an L-completely continuous nonlinear operator [9] such that

$$\|Nu\| / \|u\| \to 0 \text{ if } \|u\| \to \infty .$$

We shall be interested in the structure of the set $(\lambda,u) \in \mathbb{R} \times X$ of the solutions of the nonlinear equation

$$Lu - \lambda u = Nu \tag{4}$$

near $\lambda = 0$. Other situations can be reduced to this one by translation of λ.

THEOREM 1: Assume that 0 is an isolated eigenvalue of L with odd multiplicity and that there exists $\delta > 0$ and $R > 0$ such that each possible solution (λ,u) of (4) with $-\delta \leqq \lambda \leqq 0$ (resp. $0 \leqq \lambda \leqq \delta$) is such that $\|u\| < R$.

Then there exists $\eta > 0$ such that the following holds:

(a) equation (4) has at least one solution for $-\delta \leqq \lambda \leqq 0$ (resp. $0 \leqq \lambda \leqq \delta$);

(b) equation (4) has at least two solutions for $0 < \lambda \leqq \eta$ (resp. $-\eta \leqq \lambda < 0$).

PROOF: Dealing, say, with the first case, it is easy to show that the degree (see e.g. [9]) $D_L(L - \lambda I - N, B(R))$ of $L - \lambda I - N$ on the open ball $B(R)$ of centre zero and radius R in X is well-defined and equal to one in absolute value of all $-\delta \leqq \lambda \leqq 0$, and hence there exists $\gamma > 0$ such that the same is true for $-\delta \leqq \lambda \leqq \gamma$. Consequently, there exists a continuum C_R of solutions (λ,u) of (4) in $[-\delta,\gamma] \times B(R)$ whose projection on $\mathbb{R}$ is $[-\delta,\gamma]$. On the other hand, the results of bifurcation from infinity imply the existence of a continuum C_∞ of solutions (λ,u) of (4) bifurcating from infinity at $\lambda = 0$. More explicitly, there exists $\alpha > 0$ such that for each $0 < \varepsilon \leqq \alpha$ there is a subcontinuum $C_\varepsilon \subset C_\infty$ contained in $U_\varepsilon(0,\infty) = \{(\lambda,u) \in C_\infty : |\lambda| < \varepsilon, \|u\| > 1/\varepsilon\}$, and connecting $(0,\infty)$ to $\partial U_\varepsilon(0,\infty)$. Necessarily, for $\varepsilon = \min(1/R,\gamma,\alpha)$, we have $C_\varepsilon \subset \{(\lambda,u) \in C_\infty : 0 < \lambda < \varepsilon\}$, and hence we obtain a second solution with $\|u\| > R$ for $0 < \lambda \leqq \eta = \min(\varepsilon,\beta)$, with $\beta = \sup \{\lambda : (\lambda,u) \in C_\varepsilon\}$. □

The bifurcation diagrams are sketched on figure 2.

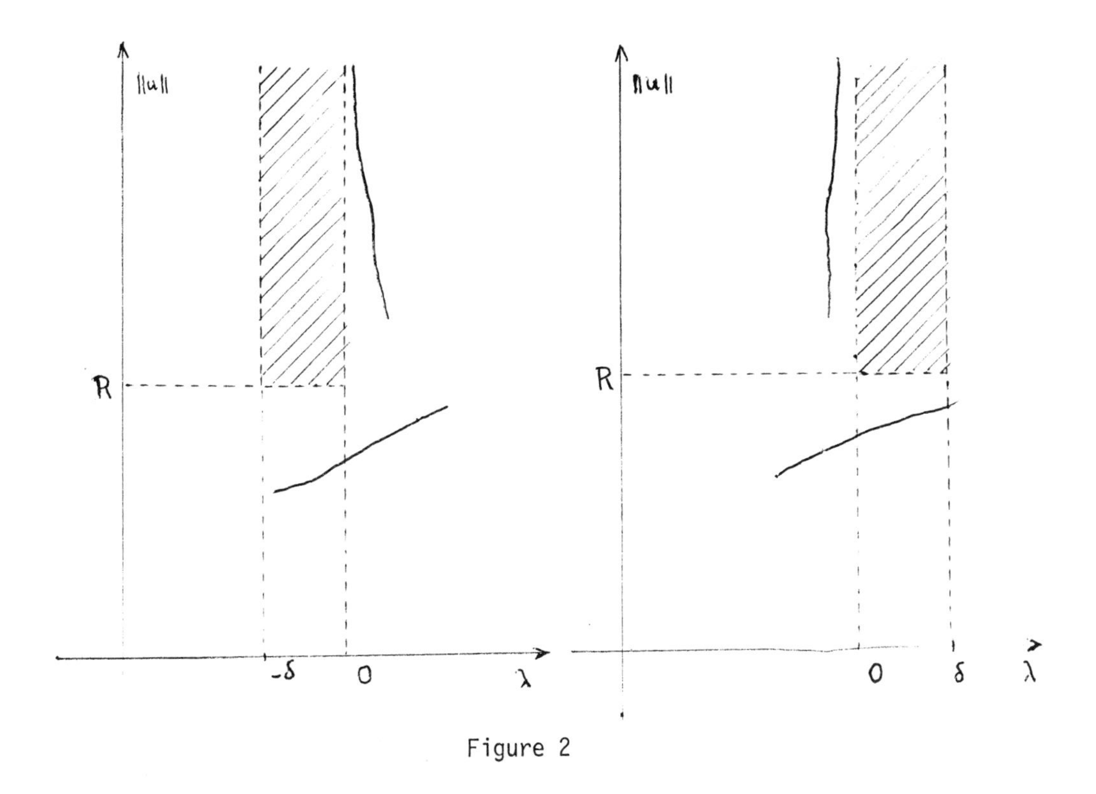

Figure 2

<u>COROLLARY 1</u>: Assume that the assumptions of Theorem 1 hold, except that the inequality sign at zero is strict. Then the conclusion (a) of Theorem 1 still holds and either (b) holds or (4) has an unbounded set of solutions for $\lambda = 0$.

<u>REMARK 1</u>: When $\lambda = 0$ is a simple eigenvalue, the results on bifurcation from infinity imply the existence of two different kinds of solutions of large norm, a positive one and a negative one (the sign being that of the projection of the solution on the normalized eigenfunction associated with the zero eigenvalue). The conclusion (b) of Theorem 1 can then be improved to the existence of at least three solutions.

Applications of those results to periodic and Dirichlet boundary value problems can be found in [10] and [11]. For example, in the case of the two-points boundary value problem

$$-u'' - n^2 u - \lambda u = g(u) - h(x),$$

$$u(0) = u(\pi) = 0$$

with $g : R \to R$ continuous and bounded, $n \in N^*$, $h \in L^2(0,\pi)$, so that we can take $X = L^2(0,\pi)$, $D(L) = H_0^2(0,\pi)$, $Lu = -u'' - n^2 u$, $Nu = g(u(\cdot))$, it has been shown in [10] and [11] that the assumptions of Theorem 1 hold in the following situations, where

$$g_\pm = \liminf_{t \to \pm\infty} g(t), \quad G_\pm = \limsup_{t \to \pm\infty} g(t),$$

(i) $n \geqq 1$, $G_- \int_0^\pi (\sin nx)^+ dx - g_+ \int_0^\pi (\sin nx)^- dx < \int_0^\pi h(x) \sin nx \, dx$

$$< g_+ \int_0^\pi (\sin nx)^+ dx - G_- \int_0^\pi (\sin nx)^- dx,$$

(ii) $n \geqq 1$, $G_+ \int_0^\pi (\sin nx)^+ dx - g_- \int_0^\pi (\sin nx)^- dx < \int_0^\pi h(x) \sin nx \, dx$

$$< g_- \int_0^\pi (\sin nx)^+ dx - G_+ \int_0^\pi (\sin nx)^- dx,$$

(iii) $n = 1$, $\int_0^\pi h(x) \sin nx \, dx = 0$ and $g(u)u > 0$ for $u \neq 0$,

(iv) $n = 1$, $\int_0^\pi h(x) \sin nx \, dx = 0$ and $g(u)u < 0$ for $u \neq 0$.

As each eigenvalue n^2 is simple, the conclusion of Remark 1 holds and if nonstrict inequalities hold in condition (iii) or (iv), Corollary 1 is applicable. In the next section, we shall apply Theorem 1 to another example.

3. A GENERALIZED STEKLOV PROBLEM

If $D = \{z \in \mathbb{C} : |z| < 1\}$ and $\Gamma = \partial D = \{z \in \mathbb{C} : z = e^{is}, s \in [0,2\pi]\}$, and if $g : R \to R$ continuous and bounded, $h \in L^2(0,2\pi)$ and $a \in R$ are given, the generalized Steklov problem [1,15] consists of finding a complex function $w = u + iv$ which is holomorphic in D, continuous on $\bar{D}$ and satisfying the conditions

$$v(0) = 0 \tag{5}$$

$$\frac{\partial u}{\partial s}(s) + av(s) = g(u(s)) - h(s), \; s \in [0,2\pi], \tag{6}$$

where, in the second equation, we write u(s) and v(s) for $u(e^{is})$ and $v(e^{is})$ respectively. If w is holomorphic on D, continuous on $\bar{D}$ and has real value u on Γ, then its imaginary part v on Γ is given by

$$v(s) = -Hu(s) + v(0)$$

where H is the Hilbert transform defined by

$$Hu(s) = (1/2\pi) \int_0^{2\pi} u(e^{it})\cot \frac{t-s}{2} \, dt.$$

Thus, in (6), because of (5), we can replace v(s) by -Hu(s). In terms of Fourier series, if

$$u(s) \sim a_0 + \sum_{k=1}^{\infty} (a_k \cos ks + b_k \sin ks),$$

then v is given by the conjugate series

$$v(s) \sim \sum_{k=1}^{\infty} (-b_k \cos ks + a_k \sin ks).$$

Therefore, if we define

$$D(L) = \{u \in L^2(\Gamma): u \text{ is absolutely continuous and } \frac{\partial u}{\partial s} \in L^2(\Gamma)\},$$

$$L : D(L) \subset L^2(\Gamma) \to L^2(\Gamma), \; u \mapsto \frac{\partial u}{\partial s} - aHu,$$

it is easy to show that L is a Fredholm linear operator of index zero with

(i) $N(L) = \text{span}(1)$ if a is not a positive integer;

(ii) $N(L) = \text{span}(1, \cos ns, \sin ns)$ if $a = n$, a positive integer;

(iii) $R(L) = \{h \in L^2(\Gamma) : \int_0^{2\pi} h(s)y(s) = 0 \text{ for each } y \in N(L)\}$.

Moreover, L has a compact resolvent and its only real eigenvalue is zero, which is simple if a is not a positive integer and has multiplicity three in the other case. Again, we set

$$g_\pm = \liminf_{t \to \pm\infty} g(t), \quad G_\pm = \limsup_{t \to \pm\infty} = \limsup_{t \quad \pm} g(t).$$

THEOREM 2: Assume that one of the following conditions hold:

(i) $a \notin \mathbb{N}^*$ and $2\pi G_- < \int_0^{2\pi} h(s)ds < 2\pi g_+$;

(ii) $a \notin \mathbb{N}^*$ and $2\pi G_+ < \int_0^{2\pi} h(s)ds < 2\pi g_-$;

(iii) $a = n \in \mathbb{N}^*$ and

$$\int_0^{2\pi} h(s)y(s)ds < \int_0^{2\pi} [g_+y^+(s) - G_-y^-(s)]ds \text{ for all } y \in N(L) \setminus \{0\};$$

(iv) $a = n \in \mathbb{N}^*$ and

$$\int_0^{2\pi} h(s)y(s)ds > \int_0^{2\pi} [G_+y^+(s) - g_-y^-(s)]ds \text{ for all } y \in N(L) \setminus \{0\}.$$

Then there exist $\delta > 0$ and $R > 0$ such that each possible solution of

$$Lu - \lambda u = g(u) - h \tag{7}$$

satisfies $\|u\|_{L^2} < R$ when $-\delta \leqq \lambda \leqq 0$ if (ii) or (iv) holds and when $0 \leqq \lambda \leqq \delta$ when (i) or (iii) holds.

PROOF: Let us consider, say, the case of condition (iii), the other ones being similar or simpler. Let u be a possible solution of (7) and let us write it $u = \bar{u} + \tilde{u}$, with $\bar{u} \in N(L)$ and $\tilde{u} \in N(L)^\perp$. If P is the orthogonal projector in $L^2(\Gamma)$ onto N(L), we deduce from (7) that

$$L\tilde{u} - \lambda\tilde{u} = (I-P)[g(u) - h],$$

and hence the boundedness of g and the compact injection of $H^1(\Gamma)$ into $C(\Gamma)$ implies the existence of $\tilde{\delta} > 0$ and $\tilde{R} > 0$ such that

$$\|\tilde{u}\|_{C(\Gamma)} < \tilde{R} \tag{8}$$

for each possible solution u of (7) with $|\lambda| \leq \tilde{\delta}$. On the other hand, we have, for each $y \in N(L)$,

$$\int_0^{2\pi} h(s)y(s)ds = \int_0^{2\pi} [g(\bar{u}(s) + \tilde{u}(s)) + \lambda u(s)]y(s)ds. \tag{9}$$

Therefore, if the conclusion of Theorem 2 does not hold, there must exist a sequence (λ_k,u_k) in $[0,\tilde{\delta}] \times L^2(\Gamma)$, satisfying (7), (8), (9) and such that $\|\bar{u}_k\| \to \infty$ if $k \to \infty$. Without loss of generality, we can also assume that

$$y_k = \bar{u}_k / \|u_k\| \to y_0 \in N(L) \cap \partial B(1)$$

uniformly on $[0,2\pi]$. Let $E_{\pm} = \{s \in [0,2\pi] : y_0(s) \gtrless 0\}$; clearly, $\|\bar{u}_k\| \; y_k(s) \to \pm \infty$ for $k \to \infty$ if $s \in E_{\pm}$ and $(0,2\pi) \setminus (E_+ \cup E_-)$ has measure zero; finally,

$$\int_0^{2\pi} y_k(s)y_0(s)ds > 0$$

for k sufficiently large. Therefore, one can deduce from (9), the above remarks and Fatou's lemma that

$$\int_0^{2\pi} h(s)y_0(s) + \int_{E_-} \limsup_{k\to\infty} g(\|\bar{u}_k\| y_k(s) + \tilde{u}_k(s))(-y_0(s))ds$$

$$\geq \int_{E_+} \liminf_{k\to\infty} g(\|\bar{u}_k\| y_k(s) + u_k(s))y_0(s)ds,$$

and hence

$$\int_0^{2\pi} h(s)y_0(s)ds \geq \int_0^{2\pi} (g_+y_0^+(s) - G_-y_0^-(s))ds,$$

a contradiction with assumption (iii). □

By combining Theorem 2 with Theorem 1 and Remark 1 (in the case where $a \notin \mathbb{N}^*$), and with Theorem 1 (when $a \in \mathbb{N}^*$), one can obtain existence and multiplicity results for (5)-(6) in the neighbourhood of $\lambda = 0$.

REFERENCES

[1] B. Ermens, Problèmes de Riemann-Hilbert non linéaires, Thèse de Doctorat, Université de Louvain, 1988.

[2] S. Fucik, Solvability of Nonlinear Equations and Boundary Value Problems, Reidel, Dordrecht, 1980.

[3] M.A. Krasnosel'skii, Eigenfunctions of nonlinear operators which approximate asymptotically to linear ones (in Russian), Dokl. Akad. Nauk SSSR, 74 (1950), 177-179.

[4] M.A. Krasnosel'skii, Topological Methods in the Theory of Nonlinear Integral Equations, Pergamon, Oxford, 1963.

[5] E.M. Landesman and A.C. Lazer, Nonlinear perturbations of linear elliptic boundary value problems at resonance, J. Math. Mech., 19 (1970), 609-623.

[6] A.C. Lazer, On Schauder's fixed point theorem and forced second order nonlinear oscillations, J. Math. Anal. Appl. 21 (1968), 421-425.

[7] A.C. Lazer and D.E. Leach, Bounded perturbations of forced harmonic oscillators at resonance, Ann. Mat. Pura Appl. (4), 82 (1969), 49-68.

[8] J. Mawhin, Landesman-Lazer's type problems for nonlinear equations, Conf. Semin. Mat. Univ. Bari, 147 (1977), 1-22.

[9] J. Mawhin, Topological Degree Methods in Nonlinear Boundary Value Problems, CBMS Reg. Conf. No. 40, American Mathematical Society, Providence, R.I., 1979.

[10] J. Mawhin and K. Schmitt, Landesman-Lazer type problems at an eigenvalue of odd multiplicity, Results Math., 14 (1988), 138-146.

[11] J. Mawhin and K. Schmitt, Nonlinear eigenvalue problems with the parameter near resonance, Ann. Polon. Math., to appear.

[12] P.H. Rabinowitz, On bifurcation from infinity, J. Differential Eqns, 14 (1973), 462-475.

[13] C.A. Stuart, Solutions of large norm for nonlinear Sturm-Liouville problems, Q. J. Math. (Oxford) (2), 24 (1973), 129-139.

[14] J.F. Toland, Asymptotic nonlinearity and nonlinear eigenvalue problems, Q.J. Math. (Oxford) (2), 24 (1973), 241-250.

[15] L. Von Wolfersdorf, Landesman-Lazer's type boundary value problems for holomorphic functions, Math. Nachr., 114 (1983), 181-189.

J. Mawhin
Université Catholique de Louvain,
Institut Mathématique,
B-1348 Louvain-la-Neuve,
Belgium

E. MEISTER AND F.-O. SPECK

Modern Wiener-Hopf methods in diffraction theory

This survey paper is an introduction to the operator theoretical approach to classes of diffraction problems for the Helmholtz equation and a half-plane screen. In contrast to the classical pioneering work of Jones, Noble, Weinstein [24,53,70] and others, problems are now studied in a well-posed Sobolev space setting. For a number of typical reference problems one-to-one correspondence to solutions of Wiener-Hopf systems is proved rigorously and leads to factorization problems for certain nonrational matrix functions. Algorithms for canonical or generalized Wiener-Hopf factorization are developed by combination of algebraic, operator and function theoretic ideas. The explicit representation of the diffracted fields can therefore be analyzed in detail, e.g. by singular expansions near the edge, which are known to be important for numerical treatment.

1. INTRODUCTION

In a famous paper [61] Sommerfeld studied in 1896 an optical diffraction problem with a semi-infinite screen, the so-called "Sommerfeld half-plane problem", which also has interpretations in acoustics and in electromagnetic theory [53]. He considered a nontransparent screen $\Sigma = \{(x,y,z) \in \mathbb{R}^3 : x > 0, y = 0\}$ and a time-harmonic incoming plane wave $\mathrm{Re}(e^{2i\pi t/\tau}\, u_{inc}(x,y,z))$ with complex amplitude

$$u_{inc}(x,y) = e^{ik(x \cos \theta + y \sin \theta)} \tag{1.1}$$

not depending upon z, i.e. the wave propagates perpendicular to the edge $x = y = 0$, $z \in \mathbb{R}$ so that we face a two-dimensional problem forgetting about z. The wave number k is assumed to fulfil $\mathrm{Re}\, k > 0$ and $\mathrm{Im}\, k > 0$ due to a lossy medium.

The diffracted or scattered field u_{inc} as well as the total field $u_{tot} = u_{inc} + u_{sc}$ then satisfy the Helmholtz equation. Further the argument

* Sponsored by the Deutsche Forschungsgemeinschaft under Grant Number KO 634/32-1.

of vanishing electrical components on the banks $\Sigma^{\pm}$ of Σ yields that the limits of u_{sc} are given by the values $g(x)$ of $-u_{inc}(x,y)$ on $\Sigma^{\pm}$; energy considerations lead to the well-known edge and radiation conditions [6,38,39]. Altogether we have for $u = u_{sc}$

$$
\begin{aligned}
(\Delta + k^2)u &= 0 && \text{in } \Omega = \mathbb{R}^2 - \bar{\Sigma} \\
u_0^{\pm} = u|_{y=\pm 0} &= g && \text{on } \Sigma^{\pm} \\
\nabla u(x,y) &= \mathcal{O}(r^{-1/2}), && r = \sqrt{(x^2 + y^2)} \to 0 \\
[(x,y)\cdot\nabla - ikr]u(x,y) &= o(r^{1/2}), && r \to \infty.
\end{aligned}
\tag{1.2}
$$

This problem can be seen as a special case from the following class of *mixed boundary-transmission problems* shown in figure 1. Those are worth studying in order to illuminate the nature of diffracted wave fields as a function of media and screen properties.

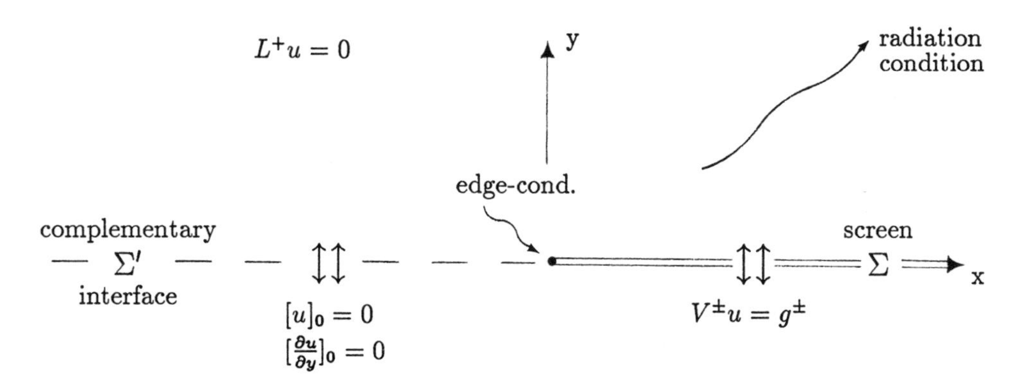

Figure 1

The Helmholtz equation is here replaced by two proper elliptic partial differential equations $L^{\pm}u = 0$ of second order with constant coefficients holding in the open half-planes $\Omega^{\pm} : y \gtrless 0$. The Dirichlet data given before on $\Sigma^\pm$ are considered as particular linear combinations $V^\pm u$ of the Cauchy data $u_0^\pm = u|_{y=\pm 0}$ and $u_1^\pm = (\partial u/\partial y)|_{y=\pm 0}$ with constant complex coefficients, which setting includes many physically relevant situations [64,65]. Since the PDE is no more assumed to hold across the "complementary screen" $\Sigma' : x < 0,\ y = 0$ we add other suitable conditions, e.g. vanishing jumps of the Cauchy data. It should be remarked that, for different media in $\Omega^\pm$, the jump conditions $[u]_0 = 0$, $[\rho\ \partial u/\partial y]_0 = 0$ with piecewise constant $\rho(x,y) = \rho^\pm$ in $\Omega^\pm$ make more sense ("refaction law" [41]). But then the analysis is not more complicated, if we even admit arbitrary boundary conditions on Σ' (from the same class as on Σ provided they are reasonable, i.e. of "normal type" [45,63]).

The desire of a well-posed setting and a general formulation of the edge and the radiation condition (which gave rise to discussions in the past [12, 20]) lead us to investigate the following principal questions.

1. The choice of appropriate function spaces (which are not proposed by nature) is influenced by: (i) the physical argument to have local finite scattering energy (edge condition) and outgoing scattered waves (radiation condition); (ii) the mathematical desire to obtain (as easy as possible) well-posed problems, i.e. existence and uniqueness of a solution and continuous dependence on the known data. This leads to Sobolev spaces, $u \in H^1(\Omega^+) \times H^1(\Omega^-)$, so that $\nabla u \in L^2$ holds (due to finite energy of the scattered field) and consequently to the trace spaces $H^{\pm 1/2} = H^{\pm 1/2}(\mathbb{R})$ for the Cauchy data $u_0^\pm$, $u_1^\pm$, to $H^{\pm 1/2}(\mathbb{R}_+)$ for their restrictions to Σ, and to the closed subspaces $\tilde{H}^{\pm 1/2}(\mathbb{R}_+)$ of $H^{\pm 1/2}$ functionals supported on $\mathbb{R}_+$ for the jumps $[u]_0 = u_0^+ - u_0^-$, $[\partial u/\partial y]_0 = u_1^+ - u_1^-$ (or, more generally, for those data which are assumed to vanish on Σ').

2. The functional analytic framework is naturally determined by the type of boundary integral equations on $\Sigma^\pm$ [45]. Systems of convolution type operators on $\mathbb{R}_+$, so-called Wiener-Hopf (WH) operators [73]

$$W : \mathop{\times}_{j=1}^{n} \tilde{H}^{r_j}(\mathbb{R}_+) \to \mathop{\times}_{j=1}^{n} H^{s_j}(\mathbb{R}_+), \tag{1.3}$$

appear in the centre of the studies ($|r_j| = |s_j| = \frac{1}{2}$). Their operator theory is best understood in the sense of pseudodifferential operators (or order $r_j - s_j$) [14,68] and general Wiener-Hopf operators [10,62].

3. The representation of the solution in the numerical sense has been frequently discussed [6,67,71] - mostly for modified geometrical situations like bounded smooth or piecewise smooth screens as obstacles. A question for effective computing (multigrid methods, mesh refinement, s - p or mixed BEM-FEMs [3]) consists of the (additive) splitting of singular terms. Local theory [14,16] tells us roughly speaking that the first term behaviour of the solution near critical points coincides with the analogue of the corresponding canonical problem (with curvilinear geometry). This motivates the investigation of half-plane problems first of all, since explicit analytical solution by means of the Fourier transformation and factorization of the symbol matrix function known from the Wiener-Hopf technique [53] yield a singular expansion of the field [14].

4. Qualitative results are supplemented by the study of the regularity of the solution in the language of H^s spaces [60] and of the far-field behaviour (radiation pattern) [25], which is also based on the explicit solution by Wiener-Hopf factorization. Since factorization methods are known in the scalar case for several classes of decomposing algebras [52] and, in the systems case for rational matrix functions [1,5], much progress has been achieved in recent years for classes of nonrational 2×2 matrix functions, which typically appear in mathematical physics and have the form

$$G = c_1 Q_1 + c_2 Q_2, \qquad (1.4)$$

with scalar functions c_j in the Wiener algebra and rational matrix functions Q_j [22,50].

We close this section with an overview in table 1 concerning the principal methods and key classification words in our philosophy for treating a reference problem P.

Table 1

Topic	Problem	Main features
Boundary value/transmission problem	Closed-form solution in a well-posed setting	Weak formulation Normal-type boundary operators Compatibility conditions
	Equivalence ⇕ Representation formulae	
Systems of WH equations	Explicit representation of bounded inverse operators	Lifting of L^2 by Bessel potential operators Fredholm criterion Modified space setting
	Equivalence ⇕ WH operator theory	
Fourier symbol matrix function	Canonical or generalized (ready) factorization	Decomposing algebras Piecewise continuous matrices $p - \rho$ regularity Partial indices
	Construction ⇑ Separation of function theoretic and algebraic aspects	
Classes of non-rational matrix functions	Factorization procedure	Rational matrix functions Commutative algebras Khrapkov and paired form matrices Rational transformation Exponential increase Pole cancellation
	Further conclusions ⇓ Fourier integrals	
Qualitative behaviour	Computation of singularities and far-field pattern	Abel-type theorems Series expansion

2. THE DIRICHLET PROBLEM

Considering the first reference problem P_D, we look for a function $u \in H^1(\Omega^+) \times H^1(\Omega^-)$, i.e. $u \in L^2(R^2)$ with $u|_{\Omega^\pm} \in H^1(\Omega^\pm)$, such that

$$(\Delta + k^2)u = 0 \qquad \text{in } \Omega^\pm \tag{2.1}$$

$$u_0^\pm = u(.,y)|_{y=\pm 0} = g^\pm \qquad \text{on } \Sigma \tag{2.2}$$

$$\left.\begin{aligned} f_0 &= u_0^+ - u_0^- = 0 \\ f_1 &= u_1^+ - u_1^- = 0 \end{aligned}\right\} \qquad \text{on } \Sigma' \tag{2.3}$$

hold where $\Omega^\pm = \{(x,y) \in R^2 : y \gtrless 0\}$ and Σ, Σ', k are defined as before, $u_1^\pm = \partial u/\partial y|_{y=\pm 0}$ denotes the unknown Neumann data (in the sense of distributions on the full line), and $g^+ = g^- \in H^{1/2}(\Sigma)$ is given (we identify Σ with $R_+ = (0,\infty)$).

We outline now the steps of the (general) solution procedure mentioned in section 1 referring to [63] for the proofs, although, in this simplest case, we will find only one scalar WH equation.

THEOREM 2.1: A function $u \in H^1(\Omega^+) \times H^1(\Omega^-)$ satisfies the Helmholtz equation (2.1), iff it is represented by

$$u(x,y) = F^{-1}_{\xi \to x}\{e^{-t(\xi)y}\hat{u}_0^+(\xi)1_+(y) + e^{t(\xi)y}\hat{u}_0^-(\xi)1_-(y)\} \tag{2.4}$$

for $(x,y) \in R^2$ with

$$\hat{u}_0^\pm(\xi) = F_{x\to\xi}\, u_0^\pm(x) = \int_{-\infty}^{+\infty} e^{ix\xi}u_0^\pm(x)dx \tag{2.5}$$

$$t(\xi) = (\xi^2 - k^2)^{1/2}, \quad \xi \in R.$$

$Fu_0^\pm$ are the Fourier transforms of the data $u_0^\pm \in H^{1/2}(R)$ (in the sense of the trace theorem [34]) and t denotes the branch of the square root that tends to $+\infty$ as $\xi \to \pm\infty$ with branch cuts along $\pm k \pm i\omega$, $\omega \geqq 0$.

For convenience we introduce the column vectors of data of a solution u of (2.1) on the line y = 0

$$u_0 = \begin{pmatrix} u_0^+ \\ u_0^- \end{pmatrix}, \quad u_1 = \begin{pmatrix} u_1^+ \\ u_1^- \end{pmatrix},$$

$$f = \begin{pmatrix} f_0 \\ f_1 \end{pmatrix}, \quad g = \begin{pmatrix} u_0^+ - u_0^- \\ u_0^+ + u_0^- \end{pmatrix}. \tag{2.6}$$

A simple but important observation follows from the representation theorem, Theorem 2.1.

<u>COROLLARY 2.2</u>: (1) The data in (2.6) satisfy

$$u_0 \in H^{1/2} \times H^{1/2}, \qquad u_1 \in H^{-1/2} \times H^{-1/2},$$

$$f \in H^{1/2} \times H^{-1/2}, \qquad g \in H^{1/2} \times H^{1/2}. \tag{2.7}$$

(2) They are in one-to-one correspondence by translation-invariant (convolution or pseudodifferential) operators on the line, in particular there holds

$$f = B_- u_0 = F^{-1} \begin{pmatrix} 1 & -1 \\ -t & -t \end{pmatrix} \cdot Fu_0$$

$$g = B_+ u_0 = F^{-1} \begin{pmatrix} 1 & -1 \\ 1 & 1 \end{pmatrix} \cdot Fu_0 \tag{2.8}$$

(with continuous "boundary operators" $B_\pm$ due to $\Sigma = \mathbb{R}_+$ and $\Sigma' = \mathbb{R}_-$).

(3) The trace operator $T_0 : u \mapsto u_0$ on the space

$$\mathcal{U} = \{u \in H^1(\Omega^+) \times H^1(\Omega^-) : (\Delta + k^2)u = 0 \text{ in } \Omega^\pm\}$$

is continuously invertible by the potential operator $G: u_0 \mapsto u$ given by the formula (2.4).

(4) For a solution of the Dirichlet problem P_D there hold

$$f \in \tilde{H}^{1/2}(\Sigma) \times \tilde{H}^{-1/2}(\Sigma), \ g \in \tilde{H}^{1/2}(\Sigma) \times H^{1/2}(\mathbb{R}) \tag{2.9}$$

i.e. $1_+ f = f$ and $1_+ g \in \tilde{H}^{1/2}(\Sigma) \times H^{1/2}(\Sigma)$ with $1_+(x) = 1$ for $x > 0$ and $1_+(x) = 0$ for $x \leq 0$.

The study of the dependence $u \leftrightarrow u_0 \leftrightarrow f \leftrightarrow g$ leads to the following equivalence theorem where we can change over to a scalar notation, since one of the (jump) data in which we represent the solution is known throughout $\Sigma \cup \Sigma'$ according to

$$g = B_+ B_-^{-1} f = F^{-1} \begin{pmatrix} 1 & 0 \\ 0 & -t^{-1} \end{pmatrix} \cdot F f. \tag{2.10}$$

THEOREM 2.3: The Dirichlet problem P_D, see (2.1)-(2.3), is equivalent to the single WH equation

$$Wf_1 = 1_+ \cdot Af_1 = -(g^+ + g^-) \tag{2.11}$$

with linear bounded operators

$$\begin{aligned} W &: \tilde{H}^{-1/2}(\Sigma) \to H^{1/2}(\Sigma) \\ A &= F^{-1} t^{-1} \cdot F : H^{-1/2} \to H^{1/2}. \end{aligned} \tag{2.12}$$

A solution f of (2.11) yields a solution u of (2.1)-(2.3) in the form

$$u = Gu_0 = GB_-^{-1} \begin{pmatrix} 0 \\ W^{-1}(-g^+ - g^-) \end{pmatrix} \tag{2.13}$$

Thus the correctness of P_D is equivalent to the bounded invertibility of W, which is known to be equivalent to a certain factorization of the Fourier symbol function t^{-1} of A [14,63,68].

THEOREM 2.4: The WH operator W in (2.11)-(2.12) is invertible by

$$W^{-1} = A_+^{-1} 1_+ \cdot A_-^{-1} \ell \tag{2.14}$$

where $\ell : H^{1/2}(\Sigma) \to H^{1/2}(\mathbb{R})$ is any extension (e.g. even extension by reflection [46]) and $A_\pm$ are defined by

$$A_\pm = F^{-1} t_\pm^{-1/2} \cdot F \ , \ t_\pm(\xi) = (\xi \pm k), \ \xi \in \mathbb{R}$$
$$A_+ : H^{-1/2} \to L^2 \ , \ A_- : L^2 \to H^{1/2} \tag{2.15}$$

(so that $t = t_- t_+$, $A = A_- A_+$ hold).

In the physically most important case where the Dirichlet data coincide on the banks of the screen ($g^+ = g^-$) and, as a consequence of (2.3), coincide also on the line ($u_0^+ = u_0^-$), the result can be written in a simpler form as follows.

<u>COROLLARY 2.5</u>: The Dirichlet problem P_D is well-posed for any $g^+ = g^- \in H^{1/2}(\Sigma)$. The solution is given by (2.13) $u = Gu_0$, $u_0 = (u_0^+, u_0^-)^T$, and

$$u_0^\pm = \Pi \ell g^+ = A_- 1_+ A_-^{-1} \ell g^+ \in H^{1/2}. \tag{2.16}$$

The dependence $g^+ \mapsto u_0^+ \mapsto u$, $H^{1/2}(\Sigma) \to H^{1/2}(\mathbb{R}) \to H^1(\mathbb{R}^2)$ is continuous.

<u>COROLLARY 2.6</u>: If $g^\pm \in H^{1/2}(\Sigma)$ differ, the Dirichlet problem is solvable iff the compatibility condition

$$g^+ - g^- \in \tilde{H}^{1/2}(\Sigma) \tag{2.17}$$

is satisfied. Then $u = Gu_0$ is given by (2.4) and

$$u_0^\pm = \pm \ \ell_0(g^+ - g^-) + \Pi \ell_e(g^+ + g^-), \tag{2.18}$$

where ℓ_0 and ℓ_e denote zero and even extension, respectively, and $\Pi = A_- 1_+ A_-^{-1}$ a projector in $H^{1/2}$ again. The dependence

$$\begin{pmatrix} g^+ - g^- \\ g^+ + g^- \end{pmatrix} \mapsto \begin{pmatrix} u_0^+ \\ u_0^- \end{pmatrix} \mapsto u \tag{2.19}$$

$$\tilde{H}^{1/2}(\Sigma) \times H^{1/2}(\Sigma) \to H^{1/2}(\mathbb{R})^2 \to H^1(\Omega^+) \times H^1(\Omega^-)$$

is continuous.

This result follows from (2.13) after replacing the zero jump by $\ell_0(g^+ - g^-)$.

A direct consequence of the solution formula is the singular behaviour near the "edge" $x = y = 0$. For simplicity consider the Neumann data jump given by

$$f_1 = -W^{-1}(g^+ + g^-) = -A_+^{-1} 1_+ A_-^{-1} \ell(g^+ + g^-) \tag{2.20}$$

in the case of smooth and rapidly decreasing (physically relevant) data $g^+ = g^- \in S(\Sigma) \subset H^{1/2}(\Sigma)$. Since f_1 does not depend on the choice of $\ell(g^+ + g^-)$ (see [14] Lemma 4.6), one can take a continuation in $S(\mathbb{R})$, which yields $\phi = A_-^{-1}\ell(g^+ + g^-) \in C^\infty(\mathbb{R}) \cap L^2(\mathbb{R})$ according to the translation invariance of A_-^{-1} and order $A_-^{-1} = 1/2$. So the singular behaviour of f_1 near the origin (as a kind of nonsmoothness) is directly connected to the action of the translation-invariant operator A_+^{-1} on functions $1_+\phi$ for $\phi \in S$ according to Abelian theorems for the Fourier transformation, i.e. it depends directly on the increase of the Fourier symbol $t_+(\xi) = (\xi + k)^{1/2}$ of A_+^{-1} at infinity [63,64]. The representation formula (4) then yields a corresponding behaviour for ∇u (whilst u is bounded).

<u>COROLLARY 2.7</u>: The solution of the Dirichlet problem for $g^\pm \in S(\Sigma)$, $g^+(+0) = g^-(+0)$, satisfies

$$\nabla u \sim \text{const}\cdot r^{-1/2}, \; r = (x^2 + y^2)^{1/2} \to 0 \text{ in } \mathbb{R}^2 - \bar{\Sigma}. \tag{2.21}$$

3. THE D/N SOMMERFELD PROBLEM

Since about 1975 several authors [19,40,56,57,63,64] investigated the problem $P_{D/N}$ given by (2.1)-(2.3) where (2.2) is replaced by

$$\begin{aligned} u_0^+ &= u(.,y)|_{y=+0} = g^+ \\ u_1^+ &= \partial u/\partial y(.,y)|_{y=-0} = g^- \end{aligned} \quad \text{on } \Sigma, \tag{3.1}$$

which are physically motivated by soft/hard-covered or perfectly conducting/nonconducting surfaces.

In our setting it is reasonable to assume $g^+ \in H^{1/2}(\Sigma)$, $g^- \in H^{-1/2}(\Sigma)$ as a consequence of Theorem 2.1. The above-mentioned procedure (as well as the classical approach) leads to a "simultaneous system" of WH equations and to the question of factoring a special matrix, which was solved by Rawlins, Heins and Meister [19,40,56,57]. But first we outline results, which are analogous to those about P_D before. The letters W, A, B_+ are used for the corresponding operators and g for data given on Σ.

THEOREM 3.1: A function $u \in H^1(\Omega^+) \times H^1(\Omega^-)$ is a solution of problem $P_{P/D}$ iff it is represented by formulae (2.4) and (2.8a) where $f = (f_0,f_1)^T \in \tilde{H}^{1/2}(\Sigma) \times \tilde{H}^{-1/2}(\Sigma)$ is a solution of

$$Wf = 1_+ Af = g \tag{3.2}$$

with (B_+ is replaced due to the boundary conditions (3.1))

$$\begin{aligned} A &= B_+B_-^{-1} = F^{-1}\Phi\cdot F : H^{1/2} \times H^{-1/2} \to H^{1/2} \times H^{-1/2} \\ \Phi &= \begin{pmatrix} 1 & 0 \\ 0 & t \end{pmatrix} \begin{pmatrix} 1 & -1 \\ -t & -t \end{pmatrix}^{-1} = -\frac{1}{2}\begin{pmatrix} -1 & t^{-1} \\ t & 1 \end{pmatrix} \\ g &= (g^+,g^-)^T \in H^{1/2}(\Sigma) \times H^{-1/2}(\Sigma). \end{aligned} \tag{3.3}$$

The problem $P_{D/N}$ is well-posed (for all data G and with respect to this topology) iff W is (bounded) invertible.

PROPOSITION 3.2: A (function theoretic) factorization of $\Phi = \tilde{\Phi}_-\tilde{\Phi}_+$ into factors, which are continuous on R, holomorphic in $\mathbb{C}^{\pm} = \{\xi \in \mathbb{C} : \text{Im } \xi \gtrless 0\}$, invertible in $\overline{\mathbb{C}^{\pm}}$, and have algebraic growth at infinity, is given by

$$\begin{pmatrix} -1 & t^{-1} \\ t & 1 \end{pmatrix} = \frac{1}{\sqrt{(4k)}} \begin{pmatrix} -t_{+-} & \frac{1}{t}t_{--} \\ tt_{--} & t_{+-} \end{pmatrix} \begin{pmatrix} t_{++} & -\frac{1}{t}t_{-+} \\ tt_{-+} & t_{++} \end{pmatrix} \tag{3.4}$$

with $t_{\pm\pm}(\xi) = [\sqrt{(2k)} \pm \sqrt{(k \pm \xi)}]^{1/2}$ where the first/second index corresponds to the first/second sign, respectively (the factors -1/2 from (3.3) and $1/\sqrt{(4k)}$ can be put somewhere).

REMARK 3.3: If the order of a translation-invariant operator T is denoted by s = ord T provided $T : H^r \to H^{r-s}$ is continuous for $r \in R$ and s is minimal, we observe (in suggestive notation for the systems case)

$$\text{ord } A = \begin{pmatrix} 0 & -1 \\ 1 & 0 \end{pmatrix} \tag{3.5}$$

$$\text{ord } \tilde{A}_{\pm}^{\pm 1} = \begin{pmatrix} 1/4 & -3/4 \\ 5/4 & 1/4 \end{pmatrix}$$

for $\tilde{A}_{\pm} = F^{-1}\tilde{\Phi}_{\pm}\cdot F$. The last assertion implies only

$$\tilde{A}_-\tilde{A}_+ : H^{1/2} \times H^{-1/2} \xrightarrow[\tilde{A}_+]{} H^{1/4} \times H^{-3/4} \xrightarrow[\tilde{A}_-]{} H^0 \times H^{-1} \tag{3.6}$$

in contrast to $A : H^{1/2} \times H^{-1/2} \to H^{1/2} \times H^{-1/2}$. Roughly speaking, all orders of the factor elements are too high by 1/4.

This phenomenon is possible, because terms of highest increase in (3.4) cancel out [19], e.g.

$$t_{+-}(\xi)t_{++}(\xi) + t_{--}(\xi)t_{-+}(\xi) \equiv \sqrt{(4k)}, \; \xi \in R \tag{3.7}$$

instead of $O(|\xi|^{1/2})$ as $|\xi| \to \infty$. As a functional analytic interpretation, a bounded operator $A : X \to X = H^{1/2} \times H^{-1/2}$ is factored into unbounded (densely defined) operators $\tilde{A}_\pm : \mathcal{D}(A_\pm) \to X$. This becomes clearer after lifting the operator W on the L^2 (order-zero) level in order to use the theory of singular integral operators of Cauchy type [52]; see [45], Prop. 3.1, for details.

LEMMA 3.4: The WH operator W in (3.2) is equivalent to the (lifted) WH operator

$$W_0 = 1_+ A_0|_{L^2(\Sigma)^2} \in L(L^2(\Sigma)^2) \tag{3.8}$$

with

$$A_0 = F^{-1}\Phi_0 \cdot F = \Lambda_- A \Lambda_+ \in L(L^2(\mathbb{R})^2) \tag{3.9}$$

$$\Phi_0 = \begin{pmatrix} t_- & 0 \\ 0 & t_-^{-1} \end{pmatrix} \Phi \begin{pmatrix} t_+^{-1} & 0 \\ 0 & t_+ \end{pmatrix} = -\frac{1}{2}\begin{pmatrix} -t_-/t_+ & 1 \\ 1 & t_+/t_- \end{pmatrix}.$$

That is, W and W_0 are connected by invertible factors:

$$\begin{aligned}
W_0 &= T_- W T_+ \\
T_+ &= 1_+ \Lambda_+ \ell_0 : L^2(\Sigma)^2 \to H^{1/2}(\Sigma) \times H^{-1/2}(\Sigma) \\
T_+^{-1} &= 1_+ \Lambda_+^{-1} \ell, \ \ell = \begin{pmatrix} \ell_e & 0 \\ 0 & \ell_{odd} \end{pmatrix} \\
T_- &= 1_+ \Lambda_- \ell : H^{1/2}(\Sigma) \times H^{-1/2}(\Sigma) \to L^2(\Sigma)^2 \\
T_-^{-1} &= 1_+ \Lambda_- \ell_0,
\end{aligned} \tag{3.10}$$

where ℓ_0, ℓ_e, ℓ_{odd} denote extensions by zero, as an even or odd functional, respectively. (ℓ_0 is usually dropped due to the identification of $L^2(\Sigma)$ and $1_+L^2(\mathbb{R})$; furthermore 1_+ can be dropped in T_+ and the extensions in $T_-^{\pm 1}$ can be replaced by any others, which act into the corresponding space $H^s(\mathbb{R})$.)

According to the holomorphy properties of $t_\pm$, WH factorizations of Φ and Φ_0 can be performed into each other. The existence of a standard [5,16] or (synonymously) a canonical [15,52] factorization of Φ_0 (which involves a bounded operator factorization of A_0 and thus of A) is disproved [54] by the fact that the symbol matrix function Φ_0 has a discontinuity at ∞ due to $t_-(\xi)/t_+(\xi) \to \pm 1$ as $\xi \to \pm\infty$. On the other hand, the theory of singular integral equations [52] ensures the existence of a *generalized factorization* (due to an unbounded operator factorization of A_0)

$$\Phi_0 = \Phi_{0-} D \Phi_{0+} \tag{3.11}$$

with factors in weighted L^2 spaces $\Phi_{0\pm}^{\pm 1} \in L^2(\mathrm{R},\rho)^{2\times 2}$, $\rho(\xi) = (\xi^2 + 1)^{-1/2}$ holomorphic extensions $\Phi_{0+}^{\pm 1}$ in $\mathbb{C}^+$ and $\Phi_{0-}^{\pm 1}$ in $\mathbb{C}^-$, and a middle term

$$D(\xi) = \mathrm{diag}\,\left(\left(\frac{\xi - i}{\xi + i}\right)^{\kappa_1},\ \left(\frac{\xi - i}{\xi + i}\right)^{\kappa_2}\right).$$

This is a consequence of piecewise continuity of Φ_0 (on $\dot{\mathrm{R}} = \mathrm{R} \cup \{\infty\}$) and the Fredholm criterion for W_0:

$$\begin{aligned} &\det \Phi_0(\xi) \neq 0,\ \xi \in \mathrm{R} \\ &\det[\mu\Phi_0(-\infty) + (1-\mu)\Phi_0(+\infty)] \neq 0,\ \mu \in [0,1]. \end{aligned} \tag{3.12}$$

A generalized factorization of Φ_0 (with D = I) can now be obtained from (3.4), (3.9) and an additional manipulation with rational matrix functions in the middle by putting

$$\begin{aligned} \Phi_0 &= \Phi_{0-} \cdot \Phi_{0+} \\ &= -\frac{1}{2} \begin{pmatrix} t_- & 0 \\ 0 & t_-^{-1} \end{pmatrix} \tilde{\Phi}_- \begin{pmatrix} 1 & 0 \\ -i\xi & 1 \end{pmatrix} \cdot \begin{pmatrix} 1 & 0 \\ i\xi & 1 \end{pmatrix} \tilde{\Phi}_+ \begin{pmatrix} t_+^{-1} & 0 \\ 0 & t_+ \end{pmatrix} \end{aligned} \tag{3.13}$$

see [64], Theorem 4.4. For the corresponding operators $A_{0\pm} = F^{-1}\Phi_{0\pm} \cdot F$ one observes the orders

$$\mathrm{ord}\, A_{0-} = \begin{pmatrix} 1/4 & -1/4 \\ 1/4 & -1/4 \end{pmatrix},\quad \mathrm{ord}\, A_{0+} = \begin{pmatrix} -1/4 & -1/4 \\ 1/4 & 1/4 \end{pmatrix} \tag{3.14}$$

which add up to zero in the composition, i.e. $A_0 = A_{0-}A_{0+}$ represents an unbounded operator factorization in the L^2 sense but the setting

$$A_{0-}A_{0+} : L^2 \times L^2 \xrightarrow[A_{0+}]{} H^{1/4} \times H^{-1/4} \xrightarrow[A_{0-}]{} L^2 \times L^2 \qquad (3.15)$$

allows an additional interpretation with bounded operators (in remarkable contrast to (3.6)) with intermediate spaces $H^{\pm 1/4}$ where the projector on $H^{\pm 1/4}(\Sigma)$ is also bounded; cf. [14]. We obtain the following results.

<u>THEOREM 3.5</u>: The WH operator W_0 in (3.8) is (bounded) invertible by

$$W_0^{-1} = A_{0+}^{-1} 1_+ \cdot A_{0-}^{-1}|_{L^2(\Sigma)^2} \qquad (3.16)$$

where $A_{0\pm}^{-1} = F^{-1}\Phi_{0\pm}^{-1} \cdot F$ are either interpreted as unbounded L^2 operators or bijections in the setting (3.15).

<u>COROLLARY 3.6</u>: The inverse of W in (3.2) is given by

$$W^{-1} = A_+^{-1} 1_+ \cdot A_-^{-1} \ell|_{H^{1/2}(\Sigma) \times H^{-1/2}(\Sigma)} \qquad (3.17)$$

$$A_\pm = F^{-1}\Phi_\pm \cdot F$$

$$\Phi_- = \tilde{\Phi}_- \begin{pmatrix} 1 & 0 \\ -i\xi & 1 \end{pmatrix}, \quad \Phi_+ = \begin{pmatrix} 1 & 0 \\ i\xi & 1 \end{pmatrix} \tilde{\Phi}_+$$

with bounded operators, if 1_+ is considered to act on $H^{1/4} \times H^{-1/4}$ (and $\ell = \mathrm{diag}(\ell_e, \ell_{odd})$ for instance).

<u>COROLLARY 3.7</u>: The solution of $P_{D/N}$ for $g \in S(\Sigma)^2 \subset H^{1/2}(\Sigma) \times H^{-1/2}(\Sigma)$ satisfies

$$\nabla u \sim \mathrm{const} \cdot r^{-3/4}, \quad r = (x^2 + y^2)^{1/2} \to 0 \text{ in } \mathbb{R}^2 - \bar{\Sigma}. \qquad (3.18)$$

4. PROBLEMS WITH GENERAL FIRST- AND SECOND-KIND TRANSMISSION CONDITIONS

In this chapter we would like to answer the question: "Is the non-square-root singular (3.18) an ordinary or an exceptional phenomenon?" For this purpose we consider a class of diffraction problems P for $u \in H^1(\Omega^+) \times H^1(\Omega^-)$ of the form

$$(\Delta + k^2)u = 0 \qquad \text{in } \Omega^\pm$$

$$\left.\begin{aligned} a_0 u_0^+ + b_0 u_0^- &= h_0 \\ a_1 u_1^+ + b_1 u_1^- &= h_1 \end{aligned}\right\} \quad \text{on } \Sigma \tag{4.2}$$

$$\left.\begin{aligned} a_0' u_0^+ + b_0' u_0^- &= h_0' \\ a_1' u_1^+ + b_1' u_1^- &= h_1' \end{aligned}\right\} \quad \text{on } \Sigma' \tag{4.3}$$

where $a_0,\ldots,b_1' \in \mathbb{C}$ are known constant coefficients and $h_0 \in H^{1/2}(\Sigma)$, $h_1 \in H^{-1/2}(\Sigma)$, $h_0' \in H^{1/2}(\Sigma')$, $h_1' \in H^{-1/2}(\Sigma')$ are assumed. There are several physically relevant examples in this class of problems; see [64] where also the case of different wave numbers in $\Omega^\pm$ is discussed.

For simplicity we avoid cases that lead to decomposing WH systems, see [63], or matrix WH operators with nonclosed ranges according to a minority of less important (artificial) diffraction problems. P is said to be of *normal type* [52], if the *boundary operators*

$$B_\pm : H^{1/2}(\mathbb{R})^2 \to H^{1/2}(\mathbb{R}) \times H^{-1/2}(\mathbb{R}) \tag{4.4}$$

$$B_+ : F^{-1}\begin{pmatrix} a_0 & b_0 \\ -a_1 t & b_1 t \end{pmatrix} \cdot F = F^{-1}\sigma_{B_+} \cdot F$$

$$B_- : F^{-1}\begin{pmatrix} a_0' & b_0' \\ -a_1' t & b_1' t \end{pmatrix} \cdot F = F^{-1}\sigma_{B_-} \cdot F$$

have Fourier symbols $\sigma_{B_\pm}$ which are *regular on* $\mathbb{R}$ and *stable at* ∞ [64], i.e.

$$\det \sigma_{B_\pm}(\xi) \neq 0, \qquad \xi \in \mathbb{R}$$
$$[\det \sigma_{B_\pm}(\xi)]^{\pm 1} = \mathcal{O}(|\xi|^{\pm 1}), \quad |\xi| \to \infty \tag{4.5}$$

hold. This obviously is equivalent to the fact that the data combinations in (4.2)-(4.3) (considered as defined on the full line) are in one-to-one correspondence with the Dirichlet data u_0 in the sense of the space setting (4.4); see Theorem 2.1 for this proof. By analogy with the previous investigation we obtain the following results [64].

THEOREM 4.1: Let P be of normal type. A function $u \in H^1(\Omega^+) \times H^1(\Omega^-)$ represents a solution of problem P, iff (i) u is of the form (2.4) where $u_0^\pm$ are given by

$$\begin{pmatrix} u_0^+ \\ u_0^- \end{pmatrix} = B_-^{-1} \left\{ \begin{pmatrix} v_+ \\ w_+ \end{pmatrix} + \begin{pmatrix} \ell_e h_0' \\ \ell_{odd} h_1' \end{pmatrix} \right\} \tag{4.6}$$

and (ii) $(v_+, w_+)^T$ is a solution of the WH system

$$W \begin{pmatrix} v_+ \\ w_+ \end{pmatrix} = \begin{pmatrix} h_0^* \\ h_1^* \end{pmatrix} = \begin{pmatrix} h_0 \\ h_1 \end{pmatrix} - 1_+ B_+ B_-^{-1} \begin{pmatrix} \ell_e h_0' \\ \ell_{odd} h_1' \end{pmatrix} \tag{4.7}$$

$$W = 1_+ \cdot F^{-1} \Phi \cdot F : \tilde{H}^{1/2}(\Sigma) \times \tilde{H}^{-1/2}(\Sigma) \to H^{1/2}(\Sigma) \times H^{-1/2}(\Sigma)$$

$$\Phi = \sigma_{B_+} \sigma_{B_-}^{-1} .$$

An elementary computation of the symbol yields

$$\Phi = \frac{1}{a_0' b_1' + b_0' a_1'} \begin{pmatrix} a_0 b_1' + b_0 a_1' & (-a_0 b_0' + b_0 a_0') t^{-1} \\ -a_1 b_1' + b_1 a_1') t & a_1 b_0' + b_1 a_0' \end{pmatrix} \tag{4.8}$$

see (3.3) for $P = P_{D/N}$. In brief and after elementary transformation this can be written as

$$\Phi = \begin{pmatrix} a & bt^{-1} \\ ct & d \end{pmatrix} \to \begin{pmatrix} 1 & t^{-1} \\ \lambda^{-1}t & 1 \end{pmatrix} = \sigma \tag{4.9}$$

iff the system does not decompose. The corresponding reference problems P_λ obviously form equivalence classes with respect to their functional analytic structure. The number

$$\lambda = \frac{ad}{bc} \in \mathbb{C} \tag{4.10}$$

is called the *characteristic parameter of P*. We introduce the lifted operators W_0, A_0 and Φ_0 by analogy to Lemma 3.4 and conclude for a normal-type problem:

PROPOSITION 4.2: W is a Fredholm operator, iff

$$ad = 0, \quad bc \neq 0 \tag{4.11}$$

or

$$ad \neq 0, \ \lambda \notin [1,\infty) \tag{4.12}$$

holds (in the decomposing or nondecomposing case, respectively).

Thus we know about the existence of a generalized factorization (3.11) of Φ_0, if (4.12) is satisfied, as we did in the case $P_{D/N}$ that corresponds to $\lambda = -1$. But the explicit factorization formulae become more complicated now. They are obtained by the method of Khrapkov [29] and Daniele [8,9]; see [64] for details.

PROPOSITION 4.3: For $\lambda \in \mathbb{C}$, $0 \neq \lambda \neq 1$, a function theoretic factorization (in the sense of Proposition 3.2) is given by

$$\sigma = \begin{pmatrix} 1 & t^{-1} \\ \lambda^{-1}t & 1 \end{pmatrix} = \sigma_-\sigma_+ \tag{4.13}$$

$$\sigma_\pm = (1-\lambda^{-1})^{1/4}\left\{\cosh[C\cdot\log\gamma_\pm]\begin{pmatrix}1 & 0\\ 0 & 1\end{pmatrix} - \sinh[C\cdot\log\gamma_\pm]\begin{pmatrix}0 & \lambda^{1/2}/t\\ t/\lambda^{1/2} & 0\end{pmatrix}\right\}$$

with

$$C = \frac{i}{\pi}\log\frac{\lambda^{1/2}+1}{\lambda^{1/2}-1} \tag{4.14}$$

$$\gamma_\pm(\xi) = \frac{(k\pm\xi)^{1/2} + i(k\mp\xi)^{1/2}}{(2k)^{1/2}}$$

and $\arg\lambda^{1/2}\in[0,\pi)$, $\arg[(\lambda^{1/2}+1)/(\lambda^{1/2}-1)]\in[-\pi,0]$, $\arg k^{1/2} = \frac{1}{2}\arg k$.

It turns out that $F^{-1}\sigma$. F suffers from a similar order deficiency as A did in (3.5) and that this can be corrected by the same trick; see [64], Chap. 4. We present the result for the reference problem P_λ, since the relation to P is obvious from (4.9).

PROPOSITION 4.4: For $\lambda \notin [1,\infty)$ a generalized factorization (3.11) of the lifted matrix is given by

$$\sigma_0 = \sigma_{0-}\cdot\sigma_{0+} = \begin{pmatrix}t_- & 0\\ 0 & t_-^{-1}\end{pmatrix}\sigma_-\begin{pmatrix}1 & 0\\ \xi/\sqrt{\lambda} & 1\end{pmatrix}\cdot\begin{pmatrix}1 & 0\\ -\xi/\sqrt{\lambda} & 1\end{pmatrix}\sigma_+\begin{pmatrix}t_+^{-1} & 0\\ 0 & t_+\end{pmatrix}. \tag{4.15}$$

The corresponding operators $A_{0\pm} = F^{-1}\sigma_{0\pm}\cdot F$ satisfy

$$\begin{aligned}\text{ord } A_{0-} &= \begin{pmatrix}\frac{1}{2}(1-\delta) & \frac{1}{2}(\delta-1)\\ \frac{1}{2}(1-\delta) & \frac{1}{2}(\delta-1)\end{pmatrix}\\ \text{ord } A_{0+} &= \begin{pmatrix}\frac{1}{2}(\delta-1) & \frac{1}{2}(\delta-1)\\ \frac{1}{2}(1-\delta) & \frac{1}{2}(1-\delta)\end{pmatrix}\end{aligned} \tag{4.16}$$

with $\delta = \operatorname{Re} C \in (0,1]$.

Therefore we obtain exactly the same interpretation as for the problem $P_{D/N}$ where $\delta = 1/2$ holds.

THEOREM 4.5: Let P be of normal type and nondecomposing ($abcd \neq 0$). The WH operator W in (4.7) is invertible (and P well-posed) iff the characteristic parameter satisfies $\lambda \notin [1,\infty)$. In this case the inverse has the form of (3.17) with

$$\Phi_- = \begin{pmatrix} 1 & 0 \\ 0 & d/b \end{pmatrix} \sigma_- \begin{pmatrix} 1 & 0 \\ \xi/\sqrt{\lambda} & 1 \end{pmatrix}$$
$$\Phi_+ = \begin{pmatrix} 1 & 0 \\ -\xi/\sqrt{\lambda} & 1 \end{pmatrix} \sigma_+ \begin{pmatrix} a & 0 \\ 0 & b \end{pmatrix} \tag{4.17}$$

(and intermediate space $H^{1/2(1-\delta)} \times H^{-1/2(1-\delta)}$).

COROLLARY 4.6: For sufficiently nice data, $(h_0^*, h_1^*) \in S(\Sigma)$ say, the singular behaviour of the solution of a problem from the class P_λ ($\lambda \notin [1,\infty)$) is described by

$$\nabla u \sim \text{const} \cdot r^{\delta/2-1}, \quad r \to 0 \text{ in } R^2 - \overline{\Sigma \cup \Sigma'} \tag{4.18}$$

with

$$\delta = \text{Re}\, \frac{i}{\pi} \log \frac{\lambda^{1/2} + 1}{\lambda^{1/2} - 1} \in (0,1].$$

The factorization (4.15) of σ_0 is bounded (and thus the related operator factors are bounded with respect to the corresponding spaces), iff $\delta = 1$, i.e. $\lambda \in (0,1)$, is satisfied. These are the only cases where the square-root singularity appears. For all other parameters $\lambda \in \mathbb{C} - [0,\infty)$ of well-posed problems, the order of the singularity is higher, namely

$$\delta/2 - 1 \in (-1, -1/2). \tag{4.19}$$

We would like to close this section with a humorous comment. The "mathematicians' answer " to our question at the beginning is that (4.19) represents the "ordinary case" whilst the square-root singularity appears only for a very small parameter set, $\lambda \in (0,1)$, of measure zero in $\mathbb{C} - [1,\infty)$, and is therefore considered "exceptional".

The "physicist's answer" might be the opposite, since most of the problems are as unnatural as $P_{D/N}$, which represents an idealized model (perfect conductance and isolation do not really exist). If one therefore accepts only impedance conditions, one finds nothing but square-root singularities, as we shall see in the next section.

5. PROBLEMS WITH IMPEDANCE AND OTHER THIRD-KIND CONDITIONS

From the physical point of view, third-kind conditions often make more sense than do Dirichlet or Neumann conditions. As a reference problem, we first study the *impedance problem* P_{Imp} with different face impedances $p^{\pm}$ on the banks of Σ where $u \in H^1(\Omega^+) \times H^1(\Omega^-)$ must satisfy

$$(\Delta + k^2)u = 0 \qquad \text{in } \Omega^{\pm} \tag{5.1}$$

$$\left.\begin{aligned} u_1^+ + ip^+u_0^+ &= h^+ \\ u_1^- - ip^-u_0^- &= h^- \end{aligned}\right\} \quad \text{on } \Sigma \tag{5.2}$$

$$\left.\begin{aligned} f_0 &= u_0^+ - u_0^- = 0 \\ f_1 &= u_1^+ - u_1^- = 0 \end{aligned}\right\} \quad \text{on } \Sigma'. \tag{5.3}$$

It is assumed that $\operatorname{Re} p^{\pm} > 0$, $\operatorname{Im} p^{\pm} > 0$ hold and $h^{\pm}$ are given in $\tilde{H}^{-1/2}(\Sigma)$.

In view of the classical approach we refer to [53,59] for $p^+ = p^-$, and to [9,21,36,37] for $p^+ \neq p^-$. Sobolev space considerations can be found in [13, 43,65]. In our opinion the most interesting mathematical questions are: How to modify the space setting in order to obtain a well-posed problem? How to find the (function theoretic) WH factorization and to perform it into a canonical or generalized one? Therefore we start with an observation, which is similar to (2.17) [43,65].

LEMMA 5.1: A necessary condition for P_{Imp} to be solvable (in the above-mentioned setting) reads

$$h^+ - h^- \in \tilde{H}^{-1/2}(\Sigma) \tag{5.4}$$

This is a consequence of

$$h^+ - h^- = 1_+(f_1 + ip^+u_0^+ + ip^-u_0^-) \in \tilde{H}^{-1/2} + H^{1/2}(\Sigma) \tag{5.5}$$

where $H^{1/2}(\Sigma) \hookrightarrow \tilde{H}^{-1/2}(\Sigma)$ is continuously embedded (note that $\tilde{H}^{-1/2}(\Sigma)$ is a dense but nonclosed subspace of $H^{-1/2}(\Sigma)$). Before further discussion we look for the equivalent WH system, rewriting the boundary conditions on Σ in the form

$$\begin{aligned} u_1^+ - u_1^- + ip^+u_0^+ + ip^-u_0^- &= h^+ - h_- \\ u_1^+ + u_1^- + ip^+u_0^+ - ip^-u_0^- &= h^+ + h^- \end{aligned} \tag{5.6}$$

for incorporating (5.4) and getting pleasant-looking formulae, respectively.

PROPOSITION 5.2: The impedance problem P_{Imp} is equivalent to the WH system

$$W\begin{pmatrix} f_1 \\ f_0 \end{pmatrix} = 1_+F^{-1}\Phi\cdot F\begin{pmatrix} f_1 \\ f_0 \end{pmatrix} = \begin{pmatrix} h^+ - h^- \\ -h^+ - h^- \end{pmatrix} \tag{5.7}$$

$$W : \tilde{H}^{-1/2}(\Sigma) \times \tilde{H}^{1/2}(\Sigma) \to H^{-1/2}(\Sigma)^2$$

$$\Phi = \begin{pmatrix} 1 - ipt^{-1} & iq \\ iqt^{-1} & t[1 - ipt^{-1}] \end{pmatrix}$$

with $p = \frac{1}{2}(p^+ + p^-)$, $q = \frac{1}{2}(p^+ - p^-)$.

This result is analogous to Theorem 3.1. If we continue to study the lifted operator W_0 (without compatibility condition) as in (3.9), we find

$$\begin{aligned} \Phi_0 &= \begin{pmatrix} t_-^{-1} & 0 \\ 0 & t_-^{-1} \end{pmatrix} \Phi \begin{pmatrix} t_+ & 0 \\ 0 & t_+^{-1} \end{pmatrix} \\ &= \begin{pmatrix} t_+/t_-(1-ipt^{-1}) & iqt^{-1} \\ iqt_-^{-2} & 1 - ipt^{-1} \end{pmatrix} \end{aligned} \tag{5.8}$$

$$\Phi_0(\pm\infty) = \begin{pmatrix} \pm 1 & 0 \\ 0 & 1 \end{pmatrix}.$$

This implies that W is not Fredholm, see (3.12) with $\mu = 1/2$; more precisely the range of W is not closed [52].

REMARK 5.3: In order to obtain at least a Fredholm operator $\tilde{W}$ by space modification, one has different possibilities.

(i) It seems physically most natural to incorporate only the compatibility condition (5.4) according to smooth extendability of the jump datum (5.5) and to stay in the "energy space" $H^1(R^2 - \bar{\Sigma})$ with u. This leads to the study of a more complicated modified WH operator

$$\tilde{W} : \tilde{H}^{-1/2}(\Sigma) \times \tilde{H}^{1/2}(\Sigma) \to \tilde{H}^{-1/2}(\Sigma) \times H^{-1/2}(\Sigma). \tag{5.9}$$

(ii) Following the spirit of Eskin's book [14] one may study W on the scale of Sobolev spaces, say

$$K_\alpha : \tilde{H}^{\alpha-1}(\Sigma) \times \tilde{H}^{\alpha}(\Sigma) \to H^{\alpha-1}(\Sigma)^2 \tag{5.10}$$

for $\alpha - 1/2 \notin \mathbb{Z}$ instead of $\alpha = 1/2$ [13]. This yields also regularity results and a singular expansion for the solution.

(iii) Another modification consists of the replacement of L^2 by L^p and of H^s by $W^{s,p}$, $p \in (1,\infty)$, $p \neq 2$ (p has another meaning here); see an analogous discussion in [64].

The idea of treating the WH operator in the first sense (5.9) consists of two steps. First one substitutes one component (where $H^{-1/2}(\Sigma)$ is replaced by $\tilde{H}^{-1/2}(\Sigma)$) by use of the WH operator (2.12) of the Dirichlet problem

$$W_D = 1_+ F^{-1} t^{-1} \cdot F : \tilde{H}^{-1/2}(\Sigma) \to H^{1/2}(\Sigma) \tag{5.11}$$

such that $\tilde{W}$ in (5.9) is equivalent to another modified WH operator

$$\tilde{\tilde{W}} : H^{1/2}(\Sigma) \times \tilde{H}^{1/2}(\Sigma) \to H^{1/2}(\Sigma) \times H^{-1/2}(\Sigma). \tag{5.12}$$

Secondly one turns over to a (usual) WH operator (in the sense of (1.3))

$$\bar{W} : L^2(\Sigma) \times \tilde{H}^{1/2}(\Sigma) \to L^2(\Sigma) \times H^{-1/2}(\Sigma) \tag{5.13}$$

by embedding $H^{1/2}(\Sigma) \hookrightarrow L^2(\Sigma)$. Inversion of $\bar{W}$ then yields that the subspaces in (5.12) are mapped onto each other because of invariance properties of scalar operators of the type "I+ smoothing". At the end it turns out that all the factorizations (function theoretic [36] as well as operator theoretic [13,65] versions) are connected by relatively simple transformations. We collect the results without technical details.

THEOREM 5.4: The modified impedance problem (5.1)-(5.4) is well-posed for any data $h^+ - h^- \in \tilde{H}^{-1/2}(\Sigma)$, $h^+ + h^- \in H^{-1/2}(\Sigma)$. The function u is given by formulae (2.4), substitution $u_0 = (u_0^+, u_0^-)^T \mapsto (1_+g_0, f_+)^T = (1_+(u_0^+ + u_0^-), -F^{-1}t_+ \cdot F(u_0^+ - u_0^-))$ and the unique solution of

$$\bar{W}_0 \begin{pmatrix} 1_+g_0 \\ f_+ \end{pmatrix} = 1_+ \begin{pmatrix} -F^{-1}t^{-1} \cdot F(h^+ - h^-) \\ F^{-1}t_-^{-1}\ F\ell_{odd}(h^+ + h^-) \end{pmatrix} \tag{5.14}$$

where $\bar{W}_0 = 1_+F^{-1}\bar{\Phi}_0 \cdot F : L^2(\Sigma)^2 \to L^2(\Sigma)^2$ has the Fourier symbol

$$\bar{\Phi}_0 = \begin{pmatrix} 1 - ipt^{-1} & iqt^{-1}t_+^{-1} \\ iqt_-^{-1} & 1 - ipt^{-1} \end{pmatrix} \tag{5.15}$$

which admits a bounded factorization (3.11) $\bar{\Phi}_0 = \bar{\Phi}_{0-}\ \bar{\Phi}_{0+}$, $\bar{\Phi}_{0\pm} \in C(\dot{R})^{2\times 2}$, i.e. there holds

$$\bar{W}_0^{-1} = F^{-1}\bar{\Phi}_{0+}^{-1} \cdot F1_+ \cdot F^{-1}\bar{\Phi}_0^{-1} \cdot F|_{L^2(\Sigma)^2} \ . \tag{5.16}$$

As mentioned above, this implies

$$\nabla u \sim \text{const} \cdot r^{-1/2}, \quad r \to 0 \text{ in } \mathbb{R}^2 - \bar{\Sigma}. \tag{5.17}$$

The method to obtain $\bar{\Phi}_{0\pm}$ explicitly [13] is similar to the procedure in section 4. We shall come back to it later in a general context.

Other relevant boundary/transmission conditions like reactance conditions [31]

$$\begin{aligned} u_0^+ - u_0^- &= 0 \\ u_1^+ - u_1^- + \kappa u_0 &= h_1 \in H^{-1/2}(R_+) \end{aligned} \tag{5.18}$$

and the interest in their mathematical structure motivated us to consider the following class of "screen problems" P

$$(\Delta + k^2)u = 0 \quad \text{in } \Omega^\pm \tag{5.19}$$

$$\left.\begin{aligned} a_{11}u_0^- + a_{12}u_0^+ + a_{13}u_1^- + a_{14}u_1^+ &= h_1 \\ a_{21}u_0^- + a_{22}u_0^+ + a_{23}u_1^- + a_{24}u_1^+ &= h_2 \end{aligned}\right\} \text{ on } \Sigma \tag{5.20}$$

$$\left.\begin{aligned} u_0^+ - u_0^- &= 0 \\ u_1^+ - u_1^- &= 0 \end{aligned}\right\} \text{ on } \Sigma' \tag{5.21}$$

for $u \in H^1(\Omega^+) \times H^1(\Omega^-)$ with constant coefficients and $h_j \in H^{s_j}(\Sigma)$ where $s_j = \pm 1/2$ depends on the type of the problem.

We outline the features of the operator theoretic approach (see [65] for the proofs and section 6 for explicit factoring in some of the most interesting cases). First we observe that the class splits into three subclasses according to

$$\text{rank} \begin{pmatrix} a_{13} & a_{14} \\ a_{23} & a_{24} \end{pmatrix} = 0,1,2. \tag{5.22}$$

Rank 0 corresponds to P_D and is dropped here. Rank 1 or 2 implies $(s_1,s_2) = (-1/2, +1/2)$ (after suitable linear combination of the two conditions (5.19)) or $(s_1,s_2) = (-1/2, -1/2)$ respectively. So the boundary operators read

$$B_\pm = F^{-1}\sigma_{B_\pm} \cdot F : H^{1/2} \times H^{1/2} \to H^{\pm 1/2} \times H^{\pm 1/2} \tag{5.23}$$

$$\sigma_{B_+} = \begin{pmatrix} a_{11} + a_{13}t & a_{12} - a_{14}t \\ a_{21} + a_{23}t & a_{22} - a_{24}t \end{pmatrix}, \quad \sigma_{B_-} = \begin{pmatrix} -1 & 1 \\ -t & -t \end{pmatrix}.$$

PROPOSITION 5.5: Problem P is equivalent (in the sense of Theorem 3.1) to the WH system

$$Wf = 1_+ B_+ B_-^{-1} f = h = (h_1, h_2)^T \tag{5.24}$$

$$W : \tilde{H}^{1/2}(\Sigma) \times \tilde{H}^{-1/2}(\Sigma) \to H^{-1/2}(\Sigma) \times H^{s_2}(\Sigma)$$

where the Fourier symbol matrix has the form

$$\sigma = \sigma_{B_+} \sigma_{B_-}^{-1} = \sigma_{pr} + \sigma_{sm} \tag{5.25}$$

$$= \begin{pmatrix} -\alpha_{14}t & \alpha_{13} \\ -\alpha_{24}t & \alpha_{23} \end{pmatrix} + \frac{1}{t} \begin{pmatrix} \alpha_{11}t & - \alpha_{12} \\ \alpha_{21}t & - \alpha_{22} \end{pmatrix}.$$

In this representation the coefficients $\alpha_{j\ell}$ are taken from (5.19) after reformulation in terms of the jumps and sums of the Cauchy data. The first term in (5.24) is called the principal part of σ and corresponds to the highest-order terms of (5.19) - provided we consider the case, where $\det \sigma_{pr} \neq 0$ holds. For what follows let P be of *normal type*, cf. (4.4), i.e.

$$\begin{aligned} &\det \sigma(\xi) \neq 0, \qquad \xi \in \mathbb{R} \\ &[\det \sigma(\xi)]^{\pm 1} = O(|\xi|^{\pm d}), \ \xi \to \pm \infty \end{aligned} \tag{5.26}$$

(B_- is bijective anyway). The cases d = -1,0,1 correspond to the ranks 0,1,2, respectively, in (5.21).

Let us first discuss the case d = 0. We assume $\alpha_{23} = \alpha_{24} = 0$ without loss of generality and have the six-parameter family of symbol matrices

$$\sigma = \begin{pmatrix} \alpha_{11} - \alpha_{14}t & \alpha_{13} - \alpha_{12}t^{-1} \\ \alpha_{21} & -\alpha_{22}t^{-1} \end{pmatrix}. \tag{5.27}$$

Their structrue can be discussed as in sections 3 and 4, which already contain many examples. The reactance problem (5.17) also belongs to this class. For a compact formulation of the Fredholm criterion (3.12) we abbreviate

$$\alpha_0 = \begin{pmatrix} \alpha_{11} & \alpha_{12} \\ \alpha_{21} & \alpha_{22} \end{pmatrix}, \quad \alpha_1 = \begin{pmatrix} \alpha_{13} & \alpha_{14} \\ \alpha_{23} & \alpha_{24} \end{pmatrix}, \quad \alpha_2 = \begin{pmatrix} \alpha_{13} & \alpha_{14} \\ \alpha_{22} & \alpha_{21} \end{pmatrix}. \tag{5.28}$$

<u>PROPOSITION 5.6</u>: The set of parameters $\alpha_{j\ell}$ where W is Fredholm is characterized by the following three conditions: first

$$\det \alpha_2 \neq 0, \ \alpha_{14}\alpha_{22} \neq 0 \tag{5.29}$$

secondly

$$\begin{aligned} &\det \alpha_0 = 0 \quad \text{or} \\ &\det \alpha_0 \neq 0, \ -\det\alpha_2/\det \alpha_0 \notin \Gamma = \{\zeta = t^{-1}(\xi), \ \xi \in \dot{\mathbb{R}}\} \end{aligned} \tag{5.30}$$

and thirdly

$$\begin{aligned} &\alpha_{13}\alpha_{21} = 0 \text{ or} \\ &\alpha_{13}\alpha_{21} \neq 0, \ \lambda = \frac{\alpha_{14}\alpha_{22}}{\alpha_{13}\alpha_{21}} \notin [0,1] \end{aligned}$$

where the very last condition includes (5.28).

From this result it is clear that the theory splits into many different cases, but all of them allow explicit generalized factorizations by the methods presented in section 7; see [65] for details.

The other class of problems ($d = 1$, case of rank 2, $s_2 = -1/2$) includes the impedance problem P_{Imp} and therefore has a rather subtle philosophy. For simplicifcation we premultiply the system by α_1^{-1} and obtain the four-parameter family of symbols and operators

$$\tilde{\sigma} = \alpha_1^{-1}\sigma = \begin{pmatrix} \tilde{\alpha}_{11} & 1 - \tilde{\alpha}_{12}t^{-1} \\ -t + \tilde{\alpha}_{21} & -\tilde{\alpha}_{22}t^{-1} \end{pmatrix}$$

$$\tilde{W}f = 1_+F^{-1}\tilde{\sigma}\cdot Ff = \alpha_1^{-1}h = \tilde{h} \tag{5.32}$$

$$\tilde{W} : \tilde{H}^{1/2}(\Sigma) \times \tilde{H}^{-1/2}(\Sigma) \to \tilde{H}^{-1/2}(\Sigma) \to H^{-1/2}(\Sigma).$$

The last tilde was added, since a compatibility condition

$$\tilde{h}_1 = \frac{\alpha_{24}h_1 - \alpha_{14}h_2}{\det \alpha_1} \in \tilde{H}^{-1/2}(\Sigma) \tag{5.33}$$

(instead of $h^{-1/2}(\Sigma)$) is obviously necessary for P to be solvable as it was in Lemma 5.1.

PROPOSITION 5.7: If $d = 1$ and $\det \alpha_1 \neq 0$ are satisfied, the following assertions are equivalent. (i) The operator $\tilde{W}$ defined by (5.31) is a Fredholm operator with index zero. (ii) There holds

$$\det \tilde{\sigma}(\xi) = \frac{\det \sigma(\xi)}{\det \alpha_1} \neq 0, \quad \xi \in \mathbf{R} \tag{5.34}$$

i.e. problem P is of a normal type. (iii) The characteristic numbers

$$\lambda_{\pm} = \frac{\tilde{\alpha}_{12} + \tilde{\alpha}_{21}}{2} \pm \frac{1}{2}\,[(\tilde{\alpha}_{12} - \tilde{\alpha}_{21})^2 + 4\tilde{\alpha}_{11}\tilde{\alpha}_{22}]^{1/2} \tag{5.35}$$

are not contained in the curve $\{\zeta = t(\xi),\ \xi \in \mathbf{R}\}$.

The proof [65] is based on the idea presented before Theorem 5.4 and ends with a discussion of the lifted modified symbol matrix

$$\bar{\sigma}_0 = \begin{pmatrix} \tilde{\alpha}_{11}/tt_+ & 1 - \tilde{\alpha}_{12}/t \\ \tilde{\alpha}_{21}/t - 1 & -\tilde{\alpha}_{22}/t_- \end{pmatrix}. \tag{5.36}$$

6. EXPLICIT WH FACTORIZATION OF CERTAIN NONRATIONAL 2 × 2 MATRIX FUNCTIONS

The aim of this section is to describe procedures for canonical (standard) or generalized factorization (3.11) [5,52] of 2 × 2 matrix functions of the form

$$G = c_1 Q_1 + c_2 Q_2 \tag{6.1}$$

(and related types) with scalar functions c_j in the Wiener algebra $\mathcal{W}(\mathbb{R}) = \mathbb{C} \dot{+} FL^1(\mathbb{R})$ or another decomposing or more general subalgebra of $C(\mathbb{R})$, and $Q_j \in \mathcal{R}(\mathbb{R})^{2\times 2}$, i.e. rational matrix functions without poles on $\mathbb{R}$.

All the above-mentioned lifted matrices $G = \Phi_0, \sigma_0$ are in the class of piecewise continuous matrix functions on $\dot{\mathbb{R}} = \mathbb{R} \cup \{\infty\}$ with jumps only at ∞. They were assumed to be 2-regular [52] in the sense of (3.12), which is equivalent to the existence of a generalized factorization (3.11) and to the Fredholm property of W_0. On the other hand it is not always clear that they are of the form (6.1), since two different square-roots $t_\pm(\xi) = (\xi \pm k)^{1/2}$ are involved; see e.g. (5.8).

Thus, for algebraic aspects, we also consider the unlifted Fourier symbol matrix functions $\Phi \in C(\mathbb{R})^{2\times 2}$, which are not bounded invertible on $\mathbb{R}$ in general but are always of the form (6.1), because they are rational in ξ and $t(\xi)$ (which yields $\Phi \in \mathcal{W}(\mathbb{R})^{2\times 2}$, if $\Phi(+\infty) = \Phi(-\infty)$ holds. This is one important reason for considering also the (weaker type of) *function theoretic factorization*

$$G = G_- G_+ \tag{6.2}$$

where $G_\pm^{\pm 1} \in C(\mathbb{R})^{2\times 2}$ have holomorphic extensions in $\mathbb{C}^\pm$ and either (i) no growth condition or (ii) algebraic behaviour, respectively, at infinity.

So it becomes clear that the interaction of function theoretic aspects (holomorphy, zeros and poles, asymptotics at ∞), operator theoretic features (Fredholm characteristis, order, boundedness and invariance properties of operators, compatibility conditions) and algebraic arguments (decomposing algebras, commutative matrix factorization) - besides physical relevance - yields a high grade of complexity, which means many, many cases. In order to avoid confusion we would like to describe a methodical conception and refer to original papers for examples [13,43,62,64,65]. It is our intention to obtain a *complete factorization procedure* for (6.1) by a concept which separates the above-mentioned mathematical aspects (we shall summarize our results in Conclusion 6.9; the objective of ready factorization is rather complicated; see [33]).

The central idea of commutative matrix factorization can be found in several papers; see [8,9,22,26,27,29,58,72] for instance. The most relevant work in this context was published by Khrapkov [29]; an early source is due to Heins [18] in 1950. Note that the letters f, g, $a_1, \ldots$ have a new meaning in what follows orientated by the notation in [22,50,65], which we use for reference.

A matrix $G \in C(\mathbb{R})^{2\times 2}$ is given in *Khrapkov canonical form* (K-*form*), if

$$G = a_1 I + a_2 R \tag{6.3}$$

$$R(\xi) = \begin{pmatrix} -c(\xi) & a(\xi) \\ b(\xi) & c(\xi) \end{pmatrix}, \quad \xi \in \mathbb{R}$$

holds, where I denotes the 2×2 unit matrix, $a_1, a_2 \in C(\mathbb{R})$ and a,b,c are polynomials. If $K \subset C(\mathbb{R})$ is an algebra, which contains $R(\mathbb{R})$ (e.g. an R-algebra [52]), then the set of K-form matrices (6.3) with a fixed R and $a_j \in K$ forms a *commutative* algebra $A(R;K)$ of 2×2 matrix functions.

It is easy to recognize K-form matrices by taking $a_2 R = G - \frac{1}{2}\mathrm{tr}\, G \cdot I$ with trace zero and splitting off a scalar function such that R is a polynomial matrix, e.g. of maximal order and coefficient 1 in the highest-order term of $-\det R(\xi)$ for uniqueness in the representation (6.3). In this case R is said to be the *deviator polynomial matrix* of G.

<u>THEOREM 6.1</u>: Let $G \in C(\mathbb{R})^{2\times 2}$ be a matrix function of K-form (6.3). Split

$$-\det R = c^2 + ab = g^2 f \tag{6.4}$$

into polynomials such that g^2 contains all square factors with $g(\xi) \neq 0$ for $\xi \in \mathbb{R}$. Further assume a function theoretic factorization and an additive decomposition, respectively,

$$(\det G)^{1/2} = (a_1^2 - a_2^2 g^2 f)^{1/2} = \gamma_- \gamma_+$$

$$\tau = \frac{1}{\sqrt{f}} \log \frac{a_1 + a_2 g\sqrt{f}}{a_1 - a_2 g\sqrt{f}} = \tau_- + \tau_+ \tag{6.5}$$

into functions that are holomorphic in $\mathbb{C}^{\pm}$ and continuous in $\overline{\mathbb{C}^{\pm}}$, respectively, where consistent branches are chosen. Similarly put $g = g_- g_+$ with $g_{\pm}(\xi) \neq 0$ in $\overline{\mathbb{C}_{\pm}}$. Then a function theoretic factorization of G reads

$$G = G_- G_+$$

$$G_{\pm} = \gamma_{\pm} \{\frac{g_{\mp}}{g_{\pm}} \cosh[\tfrac{1}{2} \sqrt{(f)}\tau_{\pm}]I + \frac{1}{g_{\pm}^2\sqrt{f}} \sinh[\tfrac{1}{2} \sqrt{(f)}\tau_{\pm}]R\}. \tag{6.6}$$

The question how to get from (6.1) to (6.3) was answered in [50].

LEMMA 6.2: The matrix function (6.1) can be written in K-form (6.4), iff there exist scalar rational functions q_{11}, q_{21} with

$$q_{11}Q_1 + q_{21}Q_2 = I. \tag{6.7}$$

Obviously it is easy then to obtain (6.3). Otherwise one can try to factor Q_1 (or Q_2) in order to obtain

$$G = c_1Q_1 + c_2Q_2 = Q_{1-}(c_1 I + c_2\tilde{Q})Q_{1+}$$

$$= Q_{1-}(\tilde{a}_1 I + \tilde{a}_2\tilde{R})Q_{1+} \tag{6.8}$$

where $\tilde{R} = \tilde{Q} - \frac{1}{2}\mathrm{tr}\, \tilde{Q}\cdot I$ has trace zero. This leads to the question whether the K-form can be simplified by rational transformations. In [50] we proved the following invariance property.

LEMMA 6.3: If $G = a_1 I + a_2 R$ is a K-form and transformed into (6.8), then the determinants of R and $\tilde{R}$ coincide up to a factor which is the square of a rational function.

This result can be interpreted as: f is an (algebraic) invariant under splitting off rational factors - and becomes most important by the following observation.

<u>COROLLARY 6.4</u>: If $f = 1$ and $a_1 \pm a_2 g$ admit function theoretic factorizations with algebraic behaviour at infinity, this holds also for the factors $G_\pm$, which simplify to

$$G_\pm = \frac{1}{2}\left\{\frac{g_\mp}{g_\pm}[\mu_\pm + \nu_\pm]I + \frac{1}{g_\pm^2}[\mu_\pm - \nu_\pm]R\right\} \tag{6.9}$$

$$a_1 + a_2 g = \mu = \mu_-\mu_+, \quad a_1 - a_2 g = \nu = \nu_-\nu_+.$$

<u>REMARK 6.5</u>: The question of algebraic behaviour of $G_\pm(\xi)$ for $\xi \to \pm\infty$ depends on $\tau_\pm$ in (6.5). In general we have exponential behaviour, if the degree of f is higher than 2. Daniele [9] proposed the following trick for a transformation of a function theoretic factorization $G = G_-G_+$ with exponential increase into another one $G = \tilde{G}_-\tilde{G}_+$ without. As an ansatz introduce a rational matrix function $Q = q_1 I + q_2 R \in A(R;\mathcal{R}(\mathbb{R}))$ with the same R and factor it by Khrapkov's formulae (6.6). Determine q_j such that the behaviour of the factors $Q_\pm$ coincides with the behaviour of $G_\pm$ at infinity. Then

$$G = G_-Q_-^{-1}QQ_+^{-1}G_+ \tag{6.10}$$

holds according to the commutativity of $A(R;K)$ and this can be performed into the desired factorization by factoring $Q = \tilde{Q}_-\tilde{Q}_+$ classically [5] provided it exists (which question can be answered independently, e.g. by the aim of (3.12)).

In many examples, this trick can also be used to get rid of the poles in (6.6) or (6.9), see [51], but in general not to reduce the order of algebraic increase, which may cause a diagonal middle factor in the generalized factorization (3.11) and is therefore shifted to the very end of the procedure; see [64,65] for example.

We would like to present *another* method [50] with more detailed results, which works for K-form matrices in the case $f = 1$. This approach has significant applications in elastodynamics [48,49,51] and is more general than the concept of functionally commutable matrix functions [35]. G is said to be of *paired form* [52], if it can be represented as

$$G = b_1 R_1 + b_2 R_2$$

$$b_j \in C(\mathrm{R}), \quad R_j \in R(\mathrm{R})^{2\times 2} \tag{6.11}$$

$$R_1^2 = R_1, \qquad R_2 = I - R_1.$$

<u>LEMMA 6.6:</u> For fixed R_j, the matrices (6.11) form a commutative subalgebra of $C(\mathrm{R})^{2\times 2}$.

Further they are always of K-form, see Lemma 6.2, because of

$$G = \frac{1}{2}(b_1 + b_2)I + \frac{1}{2}(b_1 - b_2)\tilde{R} = a_1 I + a_2 R \tag{6.12}$$

where $\tilde{R} = R_1 - R_2 \in R(\mathrm{R})^{2\times 2}$, $\mathrm{tr}(R_1 - R_2) = 0$. Conversely there holds

<u>LEMMA 6.7:</u> The matrix (6.1) with $c_j \in C(\mathrm{R})$ can be written in paired form, iff (6.7) is fulfilled and the corresponding deviator polynomial matrix R satisfies

$$- \det R = g^2 \tag{6.13}$$

with $g^{\pm 1} \in R(\mathrm{R})$.

<u>THEOREM 6.8:</u> Let G be of paired form (6.11) with $b_j \in W(\mathrm{R})$ (or in another decomposing algebra) and $R_1 = O(1)$ at ∞. Then a right canonical factorization of G exists, iff

$$b_j(\xi) \neq 0, \ \xi \in \dot{\mathrm{R}}, \ j = 1,2 \tag{6.14}$$

holds. It can be obtained explicitly from

$$G = (b_{1-} \ q_1 R_1 + b_{2-} R_2)\cdot\left(\frac{\zeta^{\kappa_1}}{q_1} R_1 + \frac{\zeta^{\kappa_2}}{q_2} R_2\right)\cdot(b_{1+} R_1 + b_{2+} q_2 R_2) \tag{6.15}$$

by canonical factorization of the middle factor in $R(\mathrm{R})^{2\times 2}$ where $\zeta(\xi) =$

$(\xi-i)/(\xi+i)$, κ_j and $b_{j\pm}$ are taken from the scalar factorization of b_j

$$b_j = b_{j-}\zeta^{\kappa_j} b_{j+} \tag{6.16}$$

$$b_{j\pm}(\xi) = \sqrt{(b(\infty))}\exp\{F_{x\to\xi} 1_+(x) F^{-1}_{\xi\to x} \ln[(\frac{\xi - i}{\xi + i})^{-\kappa_j} \frac{b_j(\xi)}{b_j(\infty)}]\}$$

and q_j are suitable bounded $R(R)$ elements.

CONCLUSION 6.9: For several important subclasses (loc.cit.) of 2×2 matrix functions of the form (6.1) the following procedure leads to a right canonical or generalized factorization:

(i) Transformation of G by splitting rational matrix functions (non-commutative factorization in $R(R)^{2\times 2}$) in order to get into the commutative subalgebra $A(R;K)$.

(ii) Function theoretic factorization of K-form matrices within $A(R;K)$ (by scalar factorization of the nonrational coefficients in K).

(iii) The Daniele and pole compensation trick (commutative factorization of a rational ansatz matrix within $A(R;K)$ with exponential increasing or bounded factors, respectively).

(iv) Canonical factorization of the remaining middle term matrix (non-commutative factorization in $R(R)^{2\times 2}$ again).

7. SOME RELATED WORK

Leaving the field of Sommerfeld half-line problems for the Helmholtz equation, we first think of 3D configurations with a screen Σ that represents a special Lipschitz domain [66] in the plane $\{(x,y,z) \in R^3 : y = 0\}$. Most of the above results can be extended for genuine half-plane problems where, e.g. u_{inc} in (1.1) depends on z (the Fourier transform variable of z is a fixed parameter in the WH procedure). But the function space setting $u \in H^1(\Omega^+) \times H^1(\Omega^-)$ with half-spaces $\Omega^\pm \subset R^3$ is not compatible with the increase of u_{inc} for $z \to -\infty$. This trouble does not occur for the *quarter-plane*

$$\sigma_1 = \{(x,y,z) \in R^3 : y = 0,\ x > 0,\ z > 0\}. \tag{7.1}$$

But half-plane problems with "decreasing data" in $H^{\pm 1/2}(\Sigma)$ play an auxiliary role for the solution of the quarter-plane problems [44,46]. We assume Σ to be Lipschitz in order to have a (continuous) extension operator $\ell : H^s(\Sigma) \to H^s(R^2)$, $s \in R$ [7,66]. Let $P(\Sigma,k,D)$ and $P(\Sigma,k,N)$ be the corresponding Dirichlet problem (2.1)-(2.3) and Neumann problem, respectively. They are governed by (single) WH equations

$$Wf = \chi_\Sigma \cdot Af = g$$

$$A = F^{-1} t^{\mp 1} \cdot F \tag{7.2}$$

$$W : \tilde{H}^{\mp 1/2}(\Sigma) \to H^{\pm 1/2}(\Sigma)$$

respectively. It was shown in [44] that the unique solution of $P(\Sigma,k,D)$ can be represented in the form

$$u = G\Pi\ell g \tag{7.3}$$

with an (arbitrary) extension operator $\ell : H^{1/2}(\Sigma) \to H^{1/2}(R^2)$ and the projector Π acting in $H^{1/2}$ onto $A\tilde{H}^{-1/2}(\Sigma)$ along $\tilde{H}^{1/2}(\Sigma')$. Π is orthogonal, iff $k = i$ holds, which idea leads to a series expansion of the solution of $P(\Sigma_1,i,D)$, $P(\Sigma_1,k,D)$ and $P(\Sigma_1,k,N)$ by analogy.

This simple reasoning yields straightforward results about the correctness of P and interesting relations to general WH operators [10,11,62] with insights about Babinet's principle and Bessel potential operators (unpublished work of R. Duduchava, R. Schneider and F.-O. Speck). The study of single Lipschitz or polygonal domains can also be seen as a first step for treating multi-media problems [42,55].

Another 2D configuration of traditional interest [2,17,23,28,69] consists of two parallel half-lines (plates) with shift

$$\sigma = \Sigma_0 \cup \Sigma_1$$

$$\Sigma_0 = \{(x,y) : x > 0,\ y = 0\},\ \Sigma_1 = \{(x,y) : x > b, y = a\}. \tag{7.4}$$

The (unlifted) Fourier symbol matrix functions are 4×4 sized with three nonrational entries of different behaviour (in contrast to $t_\pm$), namely

$$t(\xi) = (\xi^2 - k^2)^{1/2}, \ \varepsilon(\xi) = e^{-at(\xi)}, \ \zeta(\xi) = e^{ib\xi}, \ \xi \in \mathbb{R}. \tag{7.5}$$

A particular block structure

$$\sigma_0 = \left(\begin{array}{c|c} R_{11}(t_\pm) & \varepsilon\zeta R_{12}(t_\pm) \\ \hline \varepsilon\zeta^{-1} R_{21}(t_\pm) & R_{22}(t_\pm) \end{array}\right) \tag{7.6}$$

with 2×2 submatrices $R_{j\ell}$, which are rational in $t_\pm$, makes it possible to discuss the operator theoretic properties of P and to invert W as a perturbation of two 2×2 systems for the corresponding single plate problems by a fixed point argument [45,47].

According to the physical relevance of waveguides, see e.g. [23], it would be desirable to investigate problems with a modified space setting (non-decreasing solutions between the plates). Also N plates [28] and periodic configurations [41] are of particular interest.

Another topic with non-Khrapkov symbols are the two- or multi-media problems with different Helmholtz equations (i.e. different wave numbers k_1, k_2) in $\Omega^\pm$ [64]. In general it is impossible to transform σ elementary (i.e. by multiplication with rational matrix functions) into the form (6.1) for using commutative algebra arguments. So often the fixed point principle is used, if the assumption is reasonable that $|k_1 - k_2|$ is small [32]. An extension of the idea of paired operators (6.11) to N-part versions, cf. [4,27,40,42],

$$G = \sum_{j=1}^{N} b_j R_j \tag{7.7}$$

seems to be artificial. But in fact, there are significant applications in elastodynamics ($N = 3$) [51].

It is known that elliptic boundary value and transmission problems for PDE systems, which appear in the electromagnetic theory, elasticity and thermo-elasticity, lead to a higher number of coupled equations (up to eight even for the half-line). But a detailed study of the block structure of the symbol matrices makes it possible to reduce the systems for relevant examples [30] and to factor such complicated-looking matrix functions as in the elastodynamic case. Here one finds additional information on the following fact [51]. A decomposition into shear and pressure waves is possible but it destroys the topological simplicity of the space setting. The reason is that the div and curl operators on H^1 yield nonclosed subspaces, which phenomenon corresponds to rational transformations on the symbol level and their interpretation as unbounded operators.

ACKNOWLEDGEMENT: The authors would like to thank D.S. Jones and B.D. Sleeman for their encouragement to contribute to classical diffraction theory with an operator theoretical approach.

REFERENCES

[1] Bart, H., Gohberg, I. and Kaashoek, M.A., Minimal Factorization of Matrix and Operator Functions, Birkhäuser, Basel, 1979.

[2] Becker, M., Anwendung der Wiener-Hopf-Hilbert-Methode zur Lösung verallgemeinerter Sommerfeldscher Halbebenenprobleme, Diploma thesis, Fachbereich Mathematik, TH Darmstadt, 1982.

[3] Brebbia, C.A., Wendland, W.L. and Kuhn, G. (Eds), Boundary Elements IX, Vols. 1-3, Proc. Conf. Stuttgart 1987, Springer, Berlin, 1987.

[4] Cebotarev, G.N., On the closed form solution of the Riemann boundary value problem for a system of n pairs of functions, Uchen. Zap. Kazan. Gos. Univ., 116 (1956), 31-58.

[5] Clancey, K. and Gohberg, I., Factorization of Matrix Functions and Singular Integral Operators, Birkhäuser, Basel, 1981.

[6] Colton, D. and Kress, R., Integral Equations Methods in Scattering Theory, Wiley, New York, 1983.

[7] Costabel, M. and Stephan, E.P., A direct boundary integral equation method for transmission problems, J. Math. Anal. Appl., 106 (1985), 367-413.

[8] Daniele, V.G., On the factorization of Wiener-Hopf matrices in problems solvable with Hurd's method, IEEE Trans. Antennas and Propagation, 26 (1978), 614-616.

[9] Daniele, V.G., On the solution of two coupled Wiener-Hopf equations, SIAM J. Appl. Math., 44 (1984), 667-680.

[10] Devinatz, A. and Shinbrot, M., General Wiener-Hopf operators, Trans. AMS, 145 (1969), 467-494.

[11] dos Santos, A.F., General Wiener-Hopf operators and convolution equations on a finite interval, to appear.

[12] dos Santos, A.F. and Teixeira, F.S., The Sommerfeld problem revisited: solution spaces and the edge condition, J. Math. Anal. Appl., to appear.

[13] dos Santos, A.F., Lebre, A.B. and Teixeira, F.S., The diffraction problem for a half-plane with different face impedances revisted, J. Math. Anal. Appl., to appear.

[14] Eskin, G.I., Boundary Value Problems for Elliptic Pseudodifferential Equations, American Mathematical Society, Providence, R.I., 1981, (in Russian 1973).

[15] Gochberg, I.Z. and Feldman, I.A., Faltungsgleichungen und Projektionsverfahren zu ihrer Lösung, Birkhäuser, Basel, 1974 (in Russian 1971).

[16] Gohberg, I. and Krupnik, N., Einführung in die Theorie der eindimensionalen singulären Integraloperatoren, Birkhäuser, Basel, 1979 (in Russian 1973).

[17] Heins, A.E., The radiation and transmission properties of a pair of semi-infinite parallel plates, I, Q. Appl. Math., 6 (1948), 157-166; II, ibid., 215-220.

[18] Heins, A.E., Systems of Wiener-Hopf equations and their application to some boundary value problems in electromagnetic theory, Proc. Symp. Appl. Math., McGraw-Hill, New York, 1950, 76-81.

[19] Heins, A.E., The Sommerfeld half-plane problem revisited II: The factoring of a matrix of analytic functions, Math. Meth. Appl. Sci., 5 (1983), 14-21.

[20] Hönl, H., Maue, A.W. and Westpfahl, K., Theorie der Beugung, Handbuch der Physik 25/1, Springer, Berlin, 1961, pp. 218-573.

[21] Hurd, R.A., The Wiener-Hopf-Hilbert method for diffraction problems, Can. J. Phys., 54 (1976), 775-780.

[22] Hurd, R.A., The explicit factorization of Wiener-Hopf matrices, Preprint 1040, Fachbereich Mathematik, TH Darmstadt, 1987.

[23] Hurd, R.A. and Meister, E., Generalized waveguide bifurcation problems, Q. J. Mech. Appl. Math., 41 (1988), 127-139.

[24] Jones, D.S., A simplifying technique in the solution of a class of diffraction problems, Q. J. Math., 3 (1952), 189-196.

[25] Jones, D.S., The Theory of Electromagnetism, Pergamon, Oxford, 1964.

[26] Jones, D.S., Factorization of a Wiener-Hopf matrix, IMA J. Appl. Math., 32 (1984), 211-220.

[27] Jones, D.S., Commutative Wiener-Hopf factorization of a matrix, Proc. R. Soc. Lond. A,393 (1984), 185-192.

[28] Jones, D.S., Diffraction by three semi-infinite planes, Proc. R. Soc. Lond. A,404 (1986), 299-321.

[29] Khrapkov, A.A., Certain cases of the elastic equilibrium of an infinite wedge with a non-symmetric notch at the vertex, subjected to concentrated force, Prikl. Mat. Mekh., 35 (1971), 625-637.

[30] Kupradze, V.D. (Ed.), Three-Dimensional Problems of the Mathematical Theory of Elasticity and Thermoelasticity, North-Holland, Amsterdam, 1979 (in Russian 1976).

[31] LaHaie, I.J., Function theoretic techniques for the electromagnetic scattering by a resistive wedge, Radiation Laboratory Techn. Report 2, Univ. Michigan, Ann. Arbor, 1981.

[32] Latz, N., Wiener-Hopf-Gleichungen zu speziellen Ausbreitungsproblemen elektromagnetischer Schwingungen, Habil. thesis, TU Berlin, 1974.

[33] Lebre, A.B., Factorization in the Wiener algebra of a class of 2 × 2 matrix functions, to appear.

[34] Lions, J.L. and Magenes, E., Non-Homogeneous Boundary Value Problems and Applications I, Springer, Berlin, 1972 (in French 1968).

[35] Litvinchuk, G.S. and Spitkovskii, I.M., Factorization of Measurable Matrix Functions, Birkhäuser, Basel, 1987.

[36] Lüneburg, E. and Hurd, R.A., On the diffraction problem of a half plane with different face impedances, Can. J. Phys., 62 (1984), 853-860.

[37] Maliuzhinets, G.D., Excitation, reflection and emission of surface waves from a wedge with given face impedances, Sov. Phys. Dokl., 3 (1959), 226-268.

[38] Meister, E., Integraltransformationen mit Anwendungen auf Probleme der mathematischen Physik, Lang, Frankfurt, 1983.

[39] Meister, E., Randwertprobleme der Funktionentheorie, Teubner, Stuttgart, 1983.

[40] Meister, E., Some multiple-part Wiener-Hopf problems in mathematical physics, St. Banach Center Publ., 15 (1985), 359-407.

[41] Meister, E., Einige gelöste und ungelöste kanonische Probleme der mathematischen Beugungstheorie, Expo. Math., 5 (1987), 193-237.

[42] Meister, E. and Speck, F.-O., Some multidimensional Wiener-Hopf equations with applications, Trends Appl. Pure Math. Mech. 2, Proc. Conf. Kozubnik (Poland) 1977, Pitman, London, 1979, pp.217-262.

[43] Meister, E. and Speck, F.-O., Diffraction problems with impedance conditions, Appl. Anal., 22 (1986), 193-211.

[44] Meister, E. and Speck, F.-O., Scalar diffraction problems for Lipschitz and polygonal screens, Z. Ang. Math. Mech., 67 (1987), T434-435.

[45] Meister, E. and Speck, F.-O., Boundary integral equations methods for canonical problems in diffraction theory, In Boundary Elements IX (Eds. C.A. Brebbia, W.L. Wendland and G. Kuhn), Vol. I, Proc. Conf. Stuttgart 1987, Springer, Berlin, 1987, pp. 59-77.

[46] Meister, E. and Speck, F.-O., A contribution to the quarter-plane problem in diffraction theory, J. Math. Anal. Appl., 130 (1988), 223-236.

[47] Meister, E. and Speck, F.-O., On some generalized Sommerfeld half-plane problems, Z. Ang. Math. Mech., 68 (1988), T468-470.

[48] Meister, E. and Speck, F.-O., Elastodynamic scattering and matrix factorisation, In Elastic Wave Propagation, Proc. IUTAM Conf. Galway (Ireland) 1988, Elsevier Publ., 1989, pp. 373-380.

[49] Meister, E. and Speck, F.-O., Matrix factorization for canonical Wiener-Hopf problems in elastodynamical scattering theory, Proc. 9th Conf. Math. Phys. Karl-Marx-Stadt (GDR) 1988, to appear.

[50] Meister, E. and Speck, F.-O., Wiener-Hopf factorization of certain non-rational matrix fucntions in mathematical physics, Proc. Conf. Op. Th. Appl. Calgary (Canada) 1988, to appear.

[51] Meister, E. and Speck, F.-O., The explicit solution of elastodynamical diffraction problems by symbol factorization, Z. Anal. Anw. (1989), to appear.

[52] Mikhlin, S.G. and Prössdorf, S., Singular Integral Operators, Springer, Berlin, 1986 (in German 1980).

[53] Noble, B., Methods Based on the Wiener-Hopf Technique, Pergamon, London, 1958.

[54] Penzel, F., On the asymptotics of the solution of systems of singular integral equations with piecewise Hoelder-continuous coefficients, Asymptotic Anal., 1 (1988), 213-225.

[55] von Petersdorff, T., Boundary integral equations for mixed Dirichlet, Neumann and transmission problems, Math. Meth. Appl. Sci., to appear.

[56] Rawlins, A.D., The solution of a mixed boundary value problem in the theory of diffraction by a semi-infinite plane, Proc. R. Soc. Lond. A,346 (1975), 469-484.

[57] Rawlins, A.D., The explicit Wiener-Hopf factorization of a special matrix, Z. Ang. Math. Mech., 61 (1981), 527-528.

[58] Rawlins, A.D. and Williams, W.E., Matrix Wiener-Hopf factorisation, Q. J. Mech. Appl. Math., 34 (1981), 1-8.

[59] Senior, T.B.A., Diffraction by a semi-infinite metallic sheet, Proc. R. Soc. Edinb. A,213 (1952), 436-458.

[60] Simanca, S.R., Mixed elliptic boundary value problems, PHD thesis, Massachusetts Institute of Technology, 1985.

[61] Sommerfeld, A., Mathematische Theorie der Diffraction, Math. Ann., 47 (1896), 317-374.

[62] Speck, F.-O., General Wiener-Hopf Factorization Methods, Pitman, London, 1985.

[63] Speck, F.-O., Mixed boundary value problems of the type of Sommerfeld's half-plane problem, Proc. R. Soc. Edinb. 104,A (1986), 261-277.

[64] Speck, F.-O., Sommerfeld diffraction problems with first and second kind boundary conditions, SIAM J. Math. Anal., 20 (1989), 396-407.

[65] Speck, F.-O., Hurd, R.A. and Meister, E., Sommerfeld diffraction problems with third kind boundary conditions, SIAM J. Math. Anal., 20 (1989), to appear.

[66] Stein, E., Singular Integrals and Differentiability Properties of Functions, Princeton University Press, Princeton, 1970.

[67] Stephan, E.P., Boundary integral equations for mixed boundary value problems, screen and transmission problems in $\mathbb{R}^3$, Habilitation thesis, Fachbereich Mathematik, TH Darmstadt, 1984.

[68] Talenti, G., Sulle equazioni integrali di Wiener-Hopf, Boll. Unione Math. Ital., 7, Suppl. Fasc. 1 (1973), 18-118.

[69] Weinstein, L.A., On the theory of diffraction by two parallel half-planes, Izv. Akad. Nauk, Ser. Fiz., 12 (1944), 166-180 (in Russian).

[70] Weinstein, L.A., The Theory of Diffraction and the Factorization Method, Golem Press, Boulder, Colorado, 1969.

[71] Wendland, W., Bemerkungen zu Randelementmethoden bei Rissproblemen, In Mathematica ad diem natalem 75. E. Mohr, Universitätsbibliothek, TU Berlin, 1985, pp. 307-322.

[72] Williams, W.E., Recognition of some readily "Wiener-Hopf" factorizable matrices, IMA J. Appl. Math., 32 (1984), 367-378.

[73] Wiener, N. and Hopf, E., Über eine Klasse singulärer Integralgleichungen, Semester-Ber. Preuss. Akad. Wiss. Berlin, Phys.-Math. Kl., 30/32 (1931), 696-706.

E. Meister and F.-O. Speck
Fachbereich Mathematik
Technische Hochschule Darmstadt,
Schlossgartenstrasse 7
6100 Darmstadt
West Germany

R.A. SMITH

Orbitally stable closed trajectories of ordinary differential equations

The main aim of this lecture is to show how extended Poincaré-Bendixson theory can be used to prove the existence of an orbitally stable closed trajectory for a class of autonomous ordinary differential equations in $\mathbb{R}^n$. Consider the equation

$$\frac{dX}{dt} = F(X) \tag{1}$$

in which the function $F : S \to \mathbb{R}^n$ satisfies a local Lipschitz condition in open $S \subset \mathbb{R}^n$. Suppose that bounded open D has $\bar{D} \subset S$ and that its boundary ∂D is crossed inwards by all trajectories of (1) which meet it. Suppose furthermore that D contains only one point K such that $F(K) = 0$. To avoid technical complications we also assume that the Jacobian matrix $J(X) = \partial F/\partial X$ exists and is continuous in some neighbourhood of K with $\operatorname{Re} z \neq 0$ for all eigenvalues z of $J(K)$.

In the special case $n = 2$ the classical Poincaré-Bendixson theorem shows that if the critical point K is unstable then each trajectory in D converges to a closed trajectory as $t \to +\infty$ and D contains at least one orbitally stable closed trajectory Γ. If, in addition, $F(X)$ is analytic in $\bar{D}$ then Γ is asymptotically orbitally stable and D contains only a finite number of closed trajectories. It is well-known that these results can fail when $n > 2$ because D may contain almost-periodic trajectories or other more complicated chaotic motions. However, these results remain valid when $n > 2$ if we add the following assumption to exclude the possibility of such chaotic motions:

<u>HYPOTHESIS (H)</u>: Suppose that there exist positive constants λ, ε and a constant nonsingular real symmetric $n \times n$ matrix P with exactly two negative eigenvalues such that, for all X, Y in S,

$$(X - Y)^T P[F(X) - F(Y) + \lambda(X - Y)] \leq -\varepsilon|X - Y|^2. \tag{2}$$

The following result is proved in [7] (see also [8, section 4]):

THEOREM 1: If $n > 2$ and (H) holds then each trajectory of (1) in D converges either to K or to a closed trajectory as $t \to +\infty$. If, in addition, the unstable manifold U through K has dim U = 2 then D contains at least one orbitally stable closed trajectory Γ. If also F(X) is analytic in $\bar{D}$ then Γ is asymptotically orbitally stable and D contains only a finite number of closed trajectories.

An obvious consequence of this theorem is that D contains no chaotic motions of (1). To verify that dim U = 2 it is sufficient to show that Re $z > 0$ for exactly two eigenvalues z of J(K). Orbitally stable closed trajectories are of interest because it is only these which can be observed in practice in a physical or biological system. To apply Theorem 1 we do not need to compute the matrix P in (2); we only need to know that it exists.

We now describe a useful method for verifying (2). Consider equations (1) which have the feedback control form

$$\frac{dX}{dt} = AX + B\Phi(CX), \tag{3}$$

where A, B, C are constant real matrices of types $n \times n$, $n \times r$, $s \times n$, respectively, and $\Phi : CS \to \mathbb{R}^r$ is a C' function. Since the $r \times s$ Jacobian matrix $\Phi'(Y)$ exists for all Y in the subset CS of $\mathbb{R}^s$ we can define

$$\Lambda(CS) = \sup|\Phi'(Y)| \text{ for } Y \in CS, \tag{4}$$

where $|\cdot|$ denotes the spectral norm. If A has no eigenvalues on the line Re $z = -\lambda$ in the complex plane, we can define

$$\mu(\lambda) = \sup|C(zI - A)^{-1}B| \text{ for Re } z = -\lambda . \tag{5}$$

The following result is proved in [5, p. 702]:

FREQUENCY DOMAIN LEMMA: If $\mu(\lambda)\Lambda(CS) < 1$ and A has exactly two eigenvalues z with Re $z > -\lambda$ then there exist P, ε such that (H) holds with

$$F(X) = AX + B\Phi(CX).$$

The hypothesis on the eigenvalues of A ensures that the eigenvalues of P satisfy the requirements of (H). The spectral norm $|M|$ and the Euclidean norm $|M|_e$ of any real or complex rectangular matrix $M = (m_{ij})$ satisfy

$$[\Sigma|m_{ij}|^2]^{1/2} = |M|_e \geqq |M| = \sup [|MX|/|X|].$$

When the spectral norms in (4), (5) are replaced by Euclidean norms we get larger constants $\Lambda_e(CS)$, $\mu_e(\lambda)$ which are much easier to compute formally. Then $\mu_e(\lambda)\Lambda_e(CS) < 1$ is a sufficient condition for $\mu(\lambda)\Lambda(CS) < 1$.

To illustrate how this lemma can facilitate the application of Theorem 1 we consider the special case of the modified Michaelis-Menten equation in R^3. This is

$$\begin{aligned} dx/dt &= -x + (u + ax)y + (1 - x)bh(z), \\ dy/dt &= (x - axy - vy)c, \\ dz/dt &= (y - z)d, \end{aligned} \qquad (6)$$

in which a, b, c, d, u, v, are positive parameters and the given function $h : [0,\infty) \to (0,1]$ satisfies

$$\begin{aligned} &h(0) = 1,\ h(z) \to 0 \text{ as } z \to +\infty, \\ &0 > h'(z) \geqq -k \text{ for } 0 < z < \infty, \end{aligned} \qquad (7)$$

where k is a positive constant. This is a rescaled version of some equations arising from a yeast cell model devised by Hahn, Ortoleva and Ross [2, p.516]. Since x, y, z represent scaled chemical concentrations we will confine our attention to solutions of (6) in the positive cone R^3_+.

BOUNDEDNESS LEMMA: If $u < v$ there exists a bounded closed rectangular box $W \subset R^3_+$ such that every solution of (6) in R^3_+ ultimately enters W and remains in it thereafter. Furthermore W contains only one critical point K.

This elementary lemma is proved in [8] by considering the signs of the derivatives x', y', z' on planes parallel to the coordinate planes. For the special case when a is small and $h(z) = (1 + z^p)^{-1}$ with $p > 1$, Dai [1] used the torus principle to prove that a certain box W contains at least one closed trajectory provided that the unstable manifold U through K has dim U = 2. By adding further restrictions which include u small and p large, Dai was able to prove the existence of a unique asymptotically stable closed trajectory in W. This result cannot be used to test any numerical set of parameter values because it does not specify how large p must be nor how small a, u must be. The following more explicit result is proved in [8]:

THEOREM 2: Suppose that h(z) satisfies (7) and that

$$u < v, \; d > (3/2)(v + a) + 3[\tfrac{1}{2}(1 + b^2) + 3bck]^{1/2}. \tag{8}$$

Then each trajectory of (6) in the box W converges either to K or to a closed trajectory as $t \to +\infty$. If, in addition, the unstable manifold U through K has dim U = 2 then W contains at least one orbitally stable closed trajectory Γ. If, furthermore, h(z) is analytic in $(0,\infty)$ then Γ is asymptotically orbitally stable and W contains only a finite number of closed trajectories.

The following is a brief sketch of the proof in [8]. Equation (6) can be written in the form (3) with

$$X = \begin{vmatrix} x \\ y \\ z \end{vmatrix}, \quad A = \begin{vmatrix} 0 & 0 & 0 \\ 0 & -g & 0 \\ 0 & d & -d \end{vmatrix}, \quad B = \begin{vmatrix} \gamma & 1 \\ -c & 0 \\ 0 & 0 \end{vmatrix}, \quad C = \begin{vmatrix} p & 0 & 0 \\ 0 & q & 0 \\ 0 & 0 & r \end{vmatrix}.$$

where g, γ, p, q, r are auxiliary constants to be chosen later. An elementary calculation gives

$$\mu_e(\lambda) = \sup_{\text{Re } z = -\lambda} |C(zI - A)^{-1}B|_e, \quad \Lambda_e(C\bar{W}) = \sup_{Y \in C\bar{W}} |\Phi'(Y)|_e$$

as functions of g, γ, p, q, r. It is practicable to chose p, q, r, γ so as to minimize $\mu_e(\lambda)\Lambda_e(C\bar{W})$. We can then choose g, λ satisfying $0 < g < \lambda < d$ so that (8) implies $\mu_e(\lambda)\Lambda_e(C\bar{W}) < 1$. This ensures that the eigenvalues of A satisfy the requirement of the frequency domain lemma and that $\mu_e(\lambda)\Lambda_e(CS) < 1$ for some open $S \supset \bar{W}$. Then this lemma shows that (6) satisfies hypothesis (H) and the conclusions of Theorem 2 follow from those of Theorem 1 by choosing $D = W$. Since the size of $\Lambda_e(C\bar{W})$ depends on the size of the box W, it is important to choose W as small as is consistent with the boundedness lemma.

Theorem 2 shows that extended Poincaré-Bendixson theory can produce new information about stable closed trajectories which seems to be unobtainable by the various other methods discussed in the survey of Li [3]. In [7], Theorem 1 is used in a similar way to obtain explicit conditions for the existence of a stable closed trajectory of Rauch's equation and also of the generalized Goodwin equation.

So far we have considered only an extended version of the Poincaré-Bendixson theorem. Several related results for plane autonomous systems have been extended to higher-dimensional equations in [4,6]. To describe one of these we suppose that in (1) the Jacobian matrix $J(X) = \partial F/\partial X$ exists and is continuous throughout S. Suppose, as before, that $\bar{D} \subset S$ and ∂D is crossed inwards by all trajectories of (1), though now we allow the possibility that D may contain many critical points - even nonisolated critical points. In the special case $n = 2$, Bendixson's negative criterion shows that if D is simply connected and $0 > \text{trace } J(X)$ in D then D contains no closed trajectories and each trajectory in D converges to a critical point at $t \to +\infty$. The following generalization of this result is proved in [6, p.249]:

<u>THEOREM 3</u>: If $n > 2$ and D is simply connected then each trajectory of (1) in D converges to a critical point provided that $0 > \lambda_1(X) + \lambda_2(X)$ in $\bar{D}$, where $\lambda_1 \geq \lambda_2 \geq \dots \geq \lambda_n$ are the eigenvalues of the symmetric matrix $J(X)^T + J(X)$.

Because the formal calculation of $\lambda_1(X) + \lambda_2(X)$ is difficult it is sometimes more convenient to use instead the following corollary proved in [6, p. 253]:

<u>COROLLARY</u>: If $n > 2$ and D is simply connected then each trajectory of (1) in D converges to a critical point provided that there exists a continuous function $\theta : S \to \mathbb{R}$ and a constant real symmetric positive definite $n \times n$ matrix Q such that

$$\begin{aligned} &J(X)^T Q + QJ(X) + 2\theta(X)Q \geq 0 \text{ in } \bar{D}, \\ &(n - 2)\theta(X) + \text{trace } J(X) < 0 \text{ in } \bar{D}. \end{aligned} \tag{9}$$

Here (9) means that the symmetric matrix on its left-hand side is positive semidefinite. If we put $\theta(X) \equiv \lambda$, with constant $\lambda > 0$, we can sometimes verify (9) by means of a modified frequency domain lemma. This idea is used in [8] to obtain explicit conditions for every trajectory of (6) in the box W to converge to the critical point K as $t \to +\infty$. It is also used in [7] to obtain explicit conditions for every trajectory of the Lorenz equation to converge to a critical point.

REFERENCES

[1] Lo-sheng Dai, On the existence, uniqueness and global asymptotic stability of periodic solutions of the modified Michaelis-Menten mechanism, J. Differential Eqns, 31 (1979), 392-417.

[2] H. Hahn, P.J. Ortoleva and J. Ross, Chemical oscillations and multiple steady states due to variable boundary permeability, J. Theor. Biol., 41 (1973), 503-521.

[3] Bingxi Li, Periodic orbits of autonomous ordinary differential equations: theory and applications, Nonlinear Anal., 5 (1981), 931-958.

[4] R.A. Smith, An index theorem and Bendixson's negative criterion for certain differential equations of higher dimension, Proc. R. Soc. Edinb. A, 91 (1981), 63-77.

[5] R.A. Smith, Massera's convergence theorem for periodic nonlinear differential equations, J. Math. Anal. Appl., 120 (1986), 679-708.

[6] R.A. Smith, Some applications of Hausdorff dimension inequalities for ordinary differential equations, Proc. R. Soc. Edinb. A, 104 (1986), 235-259.

[7] R.A. Smith, Orbital stability for ordinary differential equations, J. Differential Eqns, 69 (1987), 265-287.

[8] R.A. Smith, Some modified Michaelis-Menten equations having stable closed trajectories, Proc. R. Soc. Edinb. (1988), to appear.

R.A. Smith
Department of Mathematics
University of Durham,
Durham,
U.K.

TIAN JINGHUANG

A survey of Hilbert's sixteenth problem

ABSTRACT: In this survey, we give a short history of the study of Hilbert's 16th problem. We emphasize Dulac's problem, the finiteness of the number $H_n(a,b)$, especially for the methods for quadratic systems and for higher-order systems. We pose some outstanding issues and provide the newest advances in researching the problem.

1. PROBLEM AND HISTORY

As is well-known, the so-called Hilbert's problem 16 is "the question as to the maximal number and position of Poincaré boundary cycles (limit cycles) for a differential equation of the first order and degree of the form

$$\frac{dy}{dx} = \frac{Y_n}{X_n}$$

or the equivalent system

$$\frac{dx}{dt} = \sum_{i+j=0}^{n} a_{ij}x^iy^j \equiv X_n(x,y), \quad \frac{dy}{dt} = \sum_{i+j=0}^{n} b_{ij}x^iy^j \equiv Y_n(x,y),$$

where X_n and Y_n are rational integral functions of the nth degree in x and y" [1]. Note that X_n and Y_n are relatively prime and at least one of them is of degree n. We let $E_n(a,b)$ denote the nth-degree differential system with the coefficients a_{ij}, b_{ij}, E_n the set of all systems $E_n(a,b)$ and H_n the maximal number of limit cycles of E_n respectively. Thus Hilbert's problem 16 is to find the upper bound

$$H_n = \sup \{H_n(a,b) : a_{ij} \in \mathbb{R}, b_{ij} \in \mathbb{R}\}$$

and determine the relative position of limit cycles. H_n is called the Hilbert number of E_n. The problem is the second part of the 16th of the 23 problems posed by D. Hilbert at the Second International Congress of

Mathematicians, Paris, 1900. The long history for 88 years of the study of this problem indicates that it has been the most difficult of the 23 problems. Indeed, it is still unsolved even for the simplest case $n = 2$. In two books [2,3] all Hilbert's 23 problems are mentioned except the 16th Hilbert believed that his 23 problems would have a deep significance for the development of mathematics as it entered the 20th century. The 16th problem is so. Many mathematical models from physics, engineering, biology, chemistry, economics, astrophysics and fluid mechanics are concerned with periodic solutions and limit cycles of polynomial differential systems. Since these are basic questions in the qualitative theory of ODEs, Hilbert's 16th problem has become more and more important and has attracted the attention of many pure and applied mathematicians.

2. THE FINITENESS PROBLEM

The finiteness problem of the number of limit cycles of polynomial differential systems contains the following two parts:

(i) $H_n(a,b) = \infty$ or $H_n(a,b) < \infty$ for each fixed n and given coefficients a_{ij}, b_{ij};

(ii) $H_n = \infty$ or $H_n < \infty$, for each fixed n.

The first important step in the solution of Hilbert's 16th problem is to determine whether H_n is finite for every given n. For this purpose, it must be known whether $H_n(a,b)$ is finite for a specific n and given coefficients a_{ij}, b_{ij}. Suppose that $H_n(a,b) = \infty$ for a polynomial system $E_n(a,b)$. Since a limit cycle surrounds at least a singular point and the number of singular points of any polynomial system $E_n(a,b)$ is finite, there must be at least a next of infinitely many cycles surrounding at least a common singular point. Therefore only the following four cases are possible. If these cycles are enclosed by a circle, then

(1) these cycles accumulate on a closed orbit;

(2) these cycles accumulate at a singular point;

(3) these cycles accumulate on a separatrix cycle. Otherwise,

(4) these cycles tend to infinity.

The case (1) is impossible, for the limit cycles of $E_n(a,b)$ correspond to isolated fixed points of the return map h and the return map (a two-sided analytic function) cannot have a nonisolated fixed point. In 1923, Dulac [4] claimed to have disposed of all the other cases (2), (3) and (4) and so proved that $H_n(a,b)$ (not H_n) is finite. But after 59 years, Il'yashenko [38,39] discovered an essential loophole in Dulac's proof and gave a counterexample to Dulac's lemma from which Dulac deduced, together with his theorem, the finiteness of the number of limit cycles of $E_n(a,b)$.

DULAC'S LEMMA: The germ of a semiregular map $f : (R^+,0) \to (R^+,0)$ is either an identical map or has an isolated fixed point, zero.

A so-called semiregular map $f : (R^+,0) \to (R^+,0)$ is a map defined on a semi-interval (0,b) of the positive semi-axis R^+ with the origin 0 (on a separatrix cycle or at the singular point) such that f(x) may be approximated by an asymptotic series of the form

$$cx^{v_0} + \sum_1^\infty P_j(\ln x)x^{v_j} \qquad (*)$$

(where $c > 0$, $0 < v_0 < v_1 < \dots$, $v_j \to \infty$ and P_j are polynomials) so that for any natural number N there is a partial sum $s_n(x)$ such that $f(x) - s_n(x) = o(x^N)$.

Il'yashenko's counterexample is that the map

$$f : x \to \begin{cases} x + e^{-1/x}\sin(1/x) & x > 0 \\ 0 & x = 0 \end{cases}$$

is semiregular and has a countable number of fixed points $x_k = 1/k\pi$ accumulated at zero.

The proof of Dulac for his lemma is as follows; the fixed points of the germ of f are the roots of the equation $f(x) = x$. If the principal term of the asymptotic series (*) of f is not the identical map, then the equation has an isolated zero root. If the principal term is the identical map and f is not the identical map, then the equation $f(x) = x$ is equivalent to the equation

$$P_1(\ln x)x^{v_1} + P_2(\ln x)x^{v_2} + \ldots = 0 \qquad (**)$$

where P_1 is a nonzero polynomial. The equation obtained after dividing (**) by x^{v_1} has no real roots in a sufficiently small deleted right-side neighbourhood of zero. In fact, for the first term $P_1(\ln x)$, of the reformed equation

$$\lim_{x\to 0+} P_1(\ln x) = \ell \neq 0 \text{ or } \lim_{x\to 0+} P_1(\ln x) = \infty.$$

The sum of the rest of the terms can be written as $o(x^{v_1})$ and so it tends to zero as $x \to 0^+$. Therefore the equation $f(x) = x$ has $x = 0$ as its isolated root.

A mistake in Dulac's argument above is that the equation $f(x) = x$ is not equivalent to the equation (**) if the semiregular map has the principal term x and is not the identical map and P_1 is a nonzero polynomial.

As stated above, the problem of the finiteness of the upper bound on the number of limit cycles of $E_n(a,b)$ is still unsolved and is called Dulac's finiteness problem. Under some additional conditions a few results on this problem have been obtained. In the case when $n = 2$ Chicone and Schafer [6] disposed of the cases (2) and (3) and so proved that all bounded graphs are finite and every quadratic system $E_2(a,b)$ has at most finitely many limit cycles in an arbitrary bounded domain in the plane. Using the work of Chicone and Schafer, Bamón [7] proved that all unbounded graphs are finite and therefore that every $E_2(a,b)$ has a finite number of limit cycles in the plane. So there has been an affirmative answer to Dulac's finiteness problem in the case $n = 2$. However, it is not known whether H_2 is finite. Il'yashenko [5] proved that every $E_n(a,b)$ has a finite number of limit cycles if all its (finite or infinite) singular points are not degenerate. This conclusion contains the following result of Sotomayor and Paterlini [41]: In the space of quadratic vector fields there exists an algebraic submanifold such that every field outside the submanifold has only finitely many limit cycles. Il'yashenko [42] has outlined a method for proving Dulac's finiteness theorem (i.e. $H_n(a,b) < \infty$). First, one should prove that limit cycles of an analytic vector field cannot accumulate on a separatrix cycle or at a singular point.

Secondly, one should use the following theorem of Dulac on semitransversality. A semitransversal to a separatrix cycle of an analytic vector field may be selected so that the corresponding monodromy mapping is either a plane germ, or vertical germ, or semiregular germ.

Very recently the Dulac problem might have been solved by four French mathematicians using non-convergent power series. The basic proof idea may be found in [55], [56], but the full proof is long and has not yet appeared in print (See Zentralblatt für Math.,Band 615, 58011; J.Ecall et al, Non-accumulation des cycles-limites,I,II C.R. Acad.Sci., Paris, Sér.I 304,375-377; 431-434 (1987).

Up to now nobody has found an upper bound of the number of limit cycles of a general polynomial system. But this kind of upper bound has been found for some special polynomial systems. For instance, the maximum number of limit cycles of the system

$$\dot{x} = -y + \delta x + \ell x^2 + mxy + ny^2, \ \dot{y} = x$$

or the system

$$\dot{x} = -y + \delta x + \ell x^2 + mxy + ny^2, \ \dot{y} = x(1 + by)$$

is 1. The maximum number of limit cycles of centre-symmetrical quadratic systems [53] is 2. Dilberto defined a limit cycle of $E_n(a,b)$ as strongly stable (unstable) if $\text{div}(X_n,Y_n) < 0$ (> 0) on the cycle. He proved that the sum of the number of strongly stable and strongly unstable limit cycles of $E_n(a,b)$ is smaller than $\frac{1}{2}(n - 2)(n-3) + 1$. If these cycles surround a unique singular point, then the sum is smaller than $\frac{1}{2}(n-1)$. The maximum number of limit cycles of the fifth degree system $\dot{y} = x$, $\dot{x} = y + a_0x + a_1x^3 + a_2x^5$ is two [57].

3. THE LOWER BOUND PROBLEM

It is easy to see that to find an upper bound of the number of limit cycles of a polynomial system is difficult and to find that of infinitely many polynomial systems of the same degree is much harder. On the other hand, comparatively speaking, it is easier to find a lower bound of H_n. This is because if only an example of $E_n(a,b)$ with K limit cycles were known, then it would follow that $H_n \geq K$.

The results obtained in this aspect are as follows: $H_2 \geq 4$, $H_3 \geq 5$, $H_3 \geq 11$, and

$$H_n \geq \begin{cases} \frac{1}{2}(n^2 + 5n) - 7, & \text{if } n \geq 6 \text{ is even} \\ \frac{1}{2}(n^2 + 5n) - 13, & \text{if } n \geq 9 \text{ is odd} \end{cases}$$

4. SOME METHODS OF PRODUCING LIMIT CYCLES

In order to obtain as many limit cycles as possible of $E_n(a,b)$, people generally adopt the following methods.

(a) Construct a suitable Poincaré-Bendixson annular domain (including such a kind of annular domain whose inner boundary is an asymptotically unstable or stable singular point and whose external boundary is a part of the equator). The famous example of Shi [25] with four limit cycles deals with this case.

(b) Produce limit cycles from a family or a few families of periodic cycles by a small perturbation of the system in question. (This method is used in Li's [32] example with 11 limit cycles of $E_3(a,b)$). (Poincaré bifurcation).

(c) Produce local limit cycles from a separatrix cycle with singular point(s) on it by a small perturbation. (Recently, Joyal [37] obtained a wonderful result. If a polynomial differential system has a loop with only a fine saddle of order $k - 1$ $(k > 1)$, then any perturbation of the system has at most k limit cycles and for any integer $\ell(1 \leq \ell \leq k)$, there exists a perturbation with exactly ℓ limit cycles.)

(d) Produce local limit cycles from a fine focus or a centre by small perturbations of coefficients. (Hopf bifurcation).

We will explain the methods (a), (d) and (b) respectively. In 1979 Shi [14] constructed the following quadratic system and proved the existence of four limit cycles of the system by using the method of Poincaré-Bendixson annular domains. Shi's example was

$$\begin{aligned} \dot{x} &= -10^{-200}x - 10x^2 + (5 - 10^{-13})xy + y^2 \\ \dot{y} &= x + x^2 + (-25 - 8 \times 10^{-52} + 9 \times 10^{-13})xy. \end{aligned} \qquad \text{(Shi)}$$

The system (Shi) has an unstable rough focus, $M(0,1)$ and a stable rough focus $O(0,0)$, and a straight line without contact, $1 - 25y = 0$. It has a singular point at infinity, saddle. On the Poincaré sphere there are four annular domains. Domain I is bounded by O and the contour without contact ℓ_1; domain II is bounded by ℓ_1 and the contour without contact ℓ_2; domain III is bounded by ℓ_2, the straight line without contact, and part of the equator; domain IV is bounded by M, the straight line without contact, and part of equator. The phase portrait of the system is shown in figure 1. By the Poincaré-

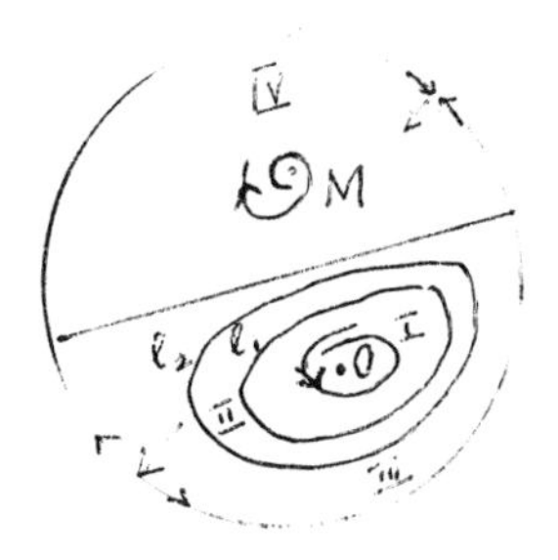

Figure 1

Bendixson annulus theorem the system (Shi) has at least four limit cycles, of which one surrounds the rough focus M and the other three surround the rough focus O.

In order to introduce the method (d), let us recall the definition of a weak (fine) focus. A focus P of the system $E_n(a,b)$ is said to be fine, if it is a centre of the corresponding linearized system

$$\dot{\xi} = \left[\left.\frac{\partial(X_n,Y_n)}{\partial(x,y)}\right|_P\right] \xi, \quad \xi \in R^2$$

where x_n, y_n are the right-hand functions of the system $E_n(a,b)$. From two foci F_1 and F_2 shown in figure 2, we can see that the focus F_2 is finer than F_1 or that F_2 is slower than F_1. We will characterize this property of foci by means of the concept "order". A fine focus of higher order is finer than one of lower order and is closer to a centre. The order of a fine focus is defined later. (Similar to focus, a limit cycle may be distinguished as rough and fine, even fine of higher order.)

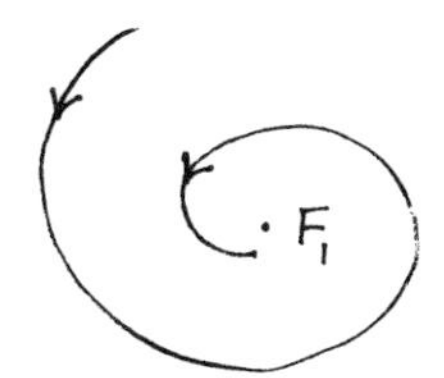

Figure 2

We start with a system having a weak (fine) focus whose order k ($k \geqq 1$) is as high as possible and perturb slightly the coefficient of the system so that the stability of the focus is changed into the opposite. Then, as a result of perturbing each time, a small-amplitude limit cycle bifurcates out of the focus. After successively perturbing the coefficients, we obtain at least k limit cycles around a rough focus. In 1952, Bautin [8] first adopted this method to obtain limit cycles of quadratic systems and proved that at most three limit cycles can appear from a centre or a fine focus with a variation of the coefficients. This method is named Bautin's technique. Since at that time there was no work on two nests of limit cycles [9], Bautin's result led to an incorrect impression that $H_2 = 3$. Hence the papers by Petrovskii and Landis [10,11] appeared early and late in 1955 and in 1957. Although the above papers had to be withdrawn because of a mistake [12], before 1979 it was still hoped that $H_2 = 3$. For example, this may be seen in the book by Ye [13] where some conclusions on various distributions of limit cycles of $E_2(a,b)$ were based on the invalid assertion of Petrovskii and Landis. In 1979, in addition to Shi [14], Chen and Wang [15] constructed quadratic systems with at least four cycles as shown in figure 3 (so $H_2 \geqq 4$).

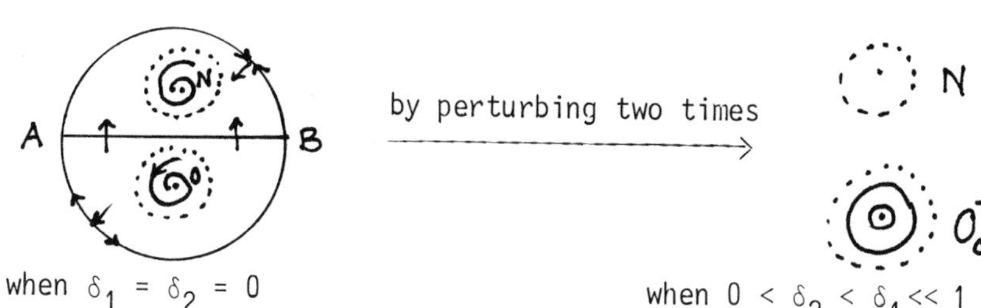

when $\delta_1 = \delta_2 = 0$

(1,1) distribution of limit cycles

N(0,1) a rough focus
O(0,0) a fine focus of order 2
AB: $1-3y = 0$ a straight line
without contact

when $0 < \delta_2 < \delta_1 << 1$

(1,3) distribution of limit cycles

N.O are fourgh foci

Figure 3

By using Bautin's technique Chen and Wang constructed the following quadratic system and proved the existence of a distribution (1,3) of limit cycles:

$$\dot{x} = -y - \delta_2 x - 3x^2 + (1 - \delta_1)xy + y^2$$
$$\dot{y} = x(1 + \tfrac{2}{9}x - 3y). \qquad \text{(Chen-Wang)}$$

In both examples of Shi and Chen-Wang, the same (1,3) configuration of limit cycles is obtained. These remarkable results have renewed interest in Hilbert's 16th problem. After that, many new authors and new results appear. Li [17], Blows and Lloyd [19], Qin, Cai and Shi [16], G. Wanner [18], Andronova [43], Cherkas [45] and Rudenok [46] found more general quadratic systems with (1,3) distribution. Andronova began with a conservative system

$$\dot{x} = -y + \ell x^2 + my^2, \quad \dot{y} = x + bxy$$

having exactly two cnetres and an infinite saddle under the conditions

$$\ell(n + b) > 0, \; n(n + b) > 0, \; n(b - \ell) < 0.$$

She constructed the system

$$\dot{x} = -y + \ell x^2 + [(2\ell+b)/(\ell+n)]axy + ny^2, \quad \dot{y} = x + ax^2 + bxy \quad (0 < a \ll 1)$$

with a rough focus at (0,1/n) and a fine focus of order at least 2 at (0,0) which approximates the conservative system. Using Bautin's technique, she proved the system has (1,3) configuration of limit cycles. In her second article [43] she further proved that in the subspace of parameters of the conservative system one picks out a region, in the neighbourhood of which there is a quadratic system with the property that the number of limit cycles is no less than four, among them no more than three around one focus at the origin and no less than one around the other. This fact is in agreement with the nonexistence of limit cycles around a fine focus of order 3 for any quadratic system [20].

Cherkas constructed a quadratic system with two parameters α, γ in the form

$$\dot{x} = (\alpha x - y)(1 + \gamma y) - \alpha x^2 + b_{11}xy + (b_{02} + \alpha a_{02})y^2$$

$$\dot{y} = (x - \alpha y)(1 + \gamma y) - x^2 - \alpha b_{11}xy + (a_{02} + \alpha b_{02})y^2$$

which has exactly a focus, (0,0) and an anti-saddle. He proved the system has at least four limit cycles with (1,3) distribution for $0 < -\alpha \ll 1$ and the corresponding γ. Dudenok proposed a method of constructing examples of the existence of at least four limit cycles for the system

$$\dot{x} = -y + \lambda x + a_{20}x^2 + 2a_{11}xy + a_{20}y^2, \quad \dot{y} = x + \lambda y + b_{20}x^2 + 2b_{11}xy + b_{02}y^2.$$

Blows and Lloyd [19] generally show how to construct quadratic systems with at least four limit cycles and their results complement the findings of many authors, such as Shi [14,47], Chen and Wang [15], Qin, Shi and Cai [48], and Li [17].

It is convenient to use Li's formulae of the focal values for a general quadratic system. The general form of quadratic systems with a fine focus at the origin is

$$\dot{x} = ax + by + \ell x^2 + mxy + ny^2, \quad \dot{y} = cx - ay + px^2 + qxy + ry^2 \quad (a^2+bc < 0).$$

Let $\sigma = (-a^2 - bc)^{1/2}$ and introduce the transformation

$$x = -\frac{1}{c}\xi - \frac{a}{\sigma}\eta, \quad y = -\frac{1}{\sigma}\eta, \quad t = \frac{c}{\sigma}\tau .$$

Then the above system is reduced to the form

$$\dot{x} = -y + a_{20}x^2 + a_{11}xy + a_{02}y^2, \quad \dot{y} = x + b_{20}x^2 + b_{11}xy + b_{02}y^2. \quad \text{(G)}$$

Extending Bautin's formulae of the focal values, Li gave the corresponding formulae for the general quadratic system (G).

THEOREM OF LI: Let

$$w_1 = A\alpha - B\beta, \quad w_2 = [\beta(5A - \beta) + \alpha(5B - \alpha)]\gamma, \quad w_3 = (A\beta + B\beta)\alpha\delta \quad \text{(W)}$$

where

$$A = a_{20} + a_{02}, \quad B = b_{20} + b_{02}, \quad \alpha = a_{11} + 2b_{02}, \quad \beta = b_{11} + 2a_{20},$$

$$\gamma = b_{20}A^3 - (a_{20}-b_{11})A^2B + (b_{02}-a_{11})AB^2 - a_{02}B^3, \quad \delta = a_{02}^2 + b_{20}^2 + a_{02}A + b_{20}B.$$

Then

(i) the origin is a fine focus of order k (k = 1,2,3) if and only if the condition (k) holds:

(1) $w_1 \neq 0$,

(2) $w_1 = 0$, $w_2 \neq 0$,

(3) $w_1 = 0$, $w_2 = 0$, $w_3 \neq 0$.

(ii) If $w_k < 0$, then the origin 0 is stable; if $w_k > 0$, then 0 is unstable.

(iii) 0 is a centre if and only if $w_1 = w_2 = w_3 = 0$.

The formulae (W) may be replaced by the following

$$w_1 = A\alpha - \beta B, \quad w_2 = \begin{cases} \beta(5A-\beta)\gamma & \text{if } A \neq 0 \\ \alpha(5B-\alpha)\gamma & \text{if } B \neq 0 \\ 0 & \text{if } A = B = 0, \end{cases} \quad w_3 = \begin{cases} A\beta\alpha\delta & \text{if } A \neq 0 \\ B\alpha\gamma\delta & \text{if } B \neq 0 \\ 0 & \text{if } A = B = 0. \end{cases}$$

<u>COROLLARY 1 OF LI'S THEOREM:</u> The origin 0 is a centre of (G) if and only if

(1) $A\alpha - B\beta = \gamma = 0$, or

(2) $\alpha = \beta = 0$, or

(3) $5A - \beta = 5B - \alpha = \delta = 0$.

<u>COROLLARY 2 OF LI'S THEOREM:</u> For Ye's form of quadratic system

$$\dot{x} = -y + \ell x^2 + mxy + ny^2, \quad \dot{y} = x + ax^2 + bxy \qquad \text{(Ye)}$$

which has a fine focus at the origin, let

$$\bar{w}_1 = m(\ell + n) - (b + 2\ell)a$$

$$\bar{w}_2 = ma(5a - m)[(\ell + n)^2(n + b) - a^2(b + 2\ell + n)]$$

$$\bar{w}_3 = ma^2[2a^2 + n(\ell + 2n)][(\ell + n)^2(n + b) - a^2(b + 2\ell + n)],$$

then

(i) (0,0) is a fine focus of order k if and only if the following condition (k) (k = 1,2,3) holds:

(1) $\bar{w}_1 \neq 0$,

(2) $\bar{w}_1 = 0, \bar{w}_2 \neq 0$,

(3) $\bar{w}_1 = \bar{w}_2 = 0, \bar{w}_3 \neq 0$.

(ii) If $\bar{w}_k < 0$ then the focus is stable; if $\bar{w}_k > 0$, then the focus is unstable.

(iii) The origin is a centre if and only if $\bar{w}_1 = \bar{w}_2 = \bar{w}_3 = 0$.

Therefore, the origin is a centre for the system (Ye) if and only if one of the following conditions (1)-(4) holds:

(1) $a = \ell + n = 0$,

(2) $m(\ell + n) = a(b + 2\ell)$, $a[(\ell + n)^2(n + b) - a^2(b + 2\ell + n)] = 0$,

(3) $m = b + 2\ell = 0$,

(4) $m = 5a$, $b = 3\ell + 5n$, $2a^2 + n(\ell + 2n) = 0$.

After having constructed a (1,3) distribution, one naturally poses the question: "Are there (0,4) distributions and/or (2,2) distributions of limit cycles of quadratic systems?" It is easily seen that a corresponding example of E_2 with (0,4) distribution would be obtained at once, if an example with a fine focus of order k (k = 1,2,3) enclosed by 4-k limit cycles were constructed. But most Chinese mathematicians in the field now believe that there can exist at most 3-k limit cycles around a fine focus of order k (k = 1,2,3) for E_2. This conjecture for k = 3 has finally been proved by Li [20]. However, it should be noted that the impossibility of (0,4) distribution cannot be derived from the conclusion just mentioned. Concerning (2,2) distribution, Ye [21] proposed a method of proving the impossibility. The present author pointed out in his talk at MIA, University of Minnesota, and at the University of Delaware (see SIAM News 13, March 1985) that workers in the field are confident that there are no more than four limit cycles for E_2, but a rigorous proof of this assertion continues to elude researchers. (Although Qin [22,44] claimed that he had proved that $H_2 = 4$, Cao [23] pointed out his problem. If Qin's work were right, then papers [6], [7], [20] and [21] mentioned above were all not necessary.)

<u>BLOWS AND LLOYD'S CONTRIBUTIONS:</u> Blows and Lloyd [19] extended Bautin's technique of creating small-amplitude limit cycles from a fine focus of $E_2(a,b)$ to higher-order systems and employed symbolic manipulation methods in computing the so-called focal values of nth-degree polynomial systems. This is a great breakthrough over Bautin's method and so it is an important contribution to Hilbert's 16th problem in the case of higher-degree systems. Therefore let us outline the ideas here.

In order to produce limit cycles from a fine focus (at the origin) the corresponding polynomial system should take the form

$$\dot{x} = y + \lambda x + \sum_{i=2}^{n} P_i(x,y)$$
$$\dot{y} = -x + \lambda y + \sum_{i=2}^{n} Q_i(x,y) \qquad (1)$$

where P_i, Q_i are homogeneous polynomials of the ith degree. As is known, a suitable Lyapunov function V for (1) should be constructed so that the rate of change along trajectories of (1) is

$$\dot{V} = n_2(x^2 + y^2) + n_4(x^2 + y^2)^2 + n_{2k}(x^2 + y^2)^k + \dots$$

where the quantities $n_2, n_4, \dots, n_{2k}, \dots$ are polynomials in the coefficients of P_i, Q_i, and λ and are called the focal values of (1). For (1), the fine focus at the origin is called of order k if $n_2 = n_4 = \dots = n_{2k} = 0$, $n_{2k+1} \neq 0$. The quantities n_2 , n_4 , ... , n_{2k} , n_{2k+2} must be computed to find the number of small-amplitude limit cycles surrounding the origin. The number k is the largest integer such that $n_2 = n_4 = \dots = n_{2k} = 0$, but all $n_{2j+2} \neq 0$ ($j = k, k + 1, \dots$). The ideal generated by the n_{2j} in the coefficient polynomials has a finite basis which consists of nonzero polynomials $n_{2k}/\langle n_2,\dots,n_{2k-2}\rangle$ where $\langle n_2,\dots,n_{2k-2}\rangle$ is the ideal generated by $n_2,\dots,n_{2k-2}$. If we write the basis as $B = \{L(0),L(1),\dots,L(M)\}$ then B is called the focal basis and the polynomials $L(0),\dots,L(M)$ are called the Lyapunov quantities of (1). Suppose the system $E_n(a,b)$ (or the system (1)) has a fine focus of order k at the origin. So $L(0) = L(1) = \dots = L(k-1) = 0$, $L(k) < 0$ (if $L(k) > 0$, the discussion is similar) and the origin is a stable fine focus of order k which is denoted by the symbol 0_k^+. Let L be a level curve of V which is close to the origin so that trajectories of (1) are inwardly across L. Now let us perturb (1) so that for the perturbed system (P_1), $L(0) = L(1) = \dots L(k-2) = 0$, but $L(k-1) > 0$. The origin is now an unstable focus which is denoted by the symbol 0_{k-1}^-. By the continuity of the vector field, the perturbation of (1) may be so small that the trajectories of (P_1) are still inwardly across L. By the Poincaré-Bendixson theorem there exists a stable limit cycle between 0_{k-1}^+ and L which is denoted by C_1^+.

Obviously, in a small enough inward neighbourhood of C_1^+ there is a closed curve C_1 such that the trajectories of the system (P_1) are outwardly across C_1. Now perturb (P_1) to (P_2) so that $L(0) = L(1) = \dots = L(k - 3) = 0$, $L(k - 2) < 0$. The origin is now a stable focus of order $k - 2$, O_{k-2}^+. By the continuity of the vector field, the perturbation of (P_1) to (P_2) may be so small that the trajectories of (P_2) are still outwardly across C_1. By the Poincare-Bendixson theorem there is an unstable limit cycle of (P_2), C_2^-, between O_{k-2}^+ and C_1. Continuing this process up to the kth perturbation, we obtain k limit cycles $C_1^+, C_2^-, \dots, C_k^+$, and the unstable rough focus O_0^-, if k is odd. Here C_1^+ is a limit cycle of the system (P_1), C_2^- is of $(P_2), \dots, C_k^+$ is of (P_K). If k is even, a stable rough focus O_0^+ is obtained after the kth perturbation of (1). Let the perturbations (P_i) to (P_{i+1}) $(i = 2,3,\dots,k-1)$ be small so that C_1 is a contour without contact of (P_k). Similarly, in an inward small neighbourhood of C_2^- there is a contour without contact of (P_k), C_2. By the Poincaré-Bendixson theorem there exists an unstable limit cycle, C_{k1}^-, between C_1 and C_2. By similar discussions, k limit cycles of the final perturbed system (P_k), $C_{k1}^-, C_{k2}^+, \dots, C_{kk-1}^-$, and C_k^+ are obtained.

Therefore, at most k small-amplitude limit cycles can bifurcate from a fine focus of order k under suitable perturbations of the polynomial system (1).

SLEEMAN'S METHOD: In his recent research report [49], based on the above work of Blows and Lloyd and using the idea of Hopf-bifurcation and the method of averaging of Show and Mallet-Paret [50], Sleeman posed a new approach to the resolution of Hilbert's 16th problem. His solution to the problem successfully depends on the manipulation of complicated algebraic quantities involving the coefficients a_{ij}, b_{ij} and leads to an algorithm for determining Lyapunov quantities and certain "averaging maps". Because of a large number of manipulations in the evaluation of integrals of trigonometric polynomials, the algorithm in the averaging process is solved using the symbolic manipulation package MACSYMA [51,52]. Finally, the results obtained are applied to quadratic and cubic systems and an outline for a systematic attack on higher-order systems is given.

CUBIC SYSTEMS: First, it is necessary and interesting to compare the known relative positions of cycles of E_2 with some relative positions of cycles of

E_3. The author [26] pointed out that many properties of cycles of E_3 differ greatly from those of E_2. For example, a closed orbit (a limit cycle) of E_3 may surround more than one critical point (as shown in figure 4); when it surrounds only a singular critical point the latter may not be a focus or a centre. (The Van der Pol equation

$$\dot{x} = y, \quad \dot{y} = -x + 3(1 - x^2)y$$

has a unique limit cycle surrounding a node.) A centre and a limit cycle of E_3 can coexist. There can exist three cycles of which one encloses the two others, which are separate, or three cycles which separate mutually. The above-mentioned cases are illustrated in figure 4. The cases (a)-(e) of limit cycles and singular points can be realized only for E_3, but not for E_2.

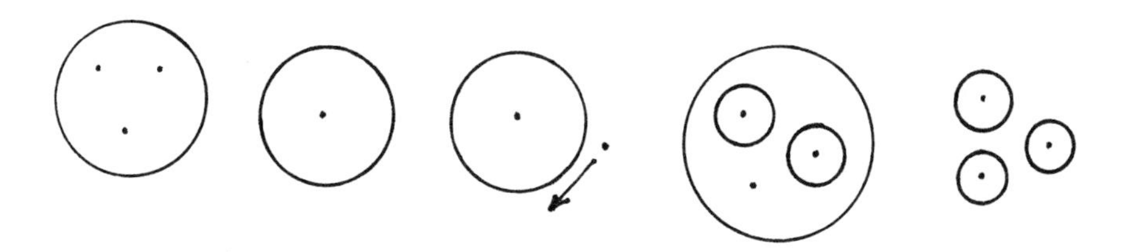

The number of singular points is 1.	a node	a centre	one surrounds two	three separate
(a)	(b)	(c)	(d)	(e)

Figure 4

Each closed orbit of any quadratic system $E_2(a,b)$ encloses a convex region, so that it is cut by any straight line in at most two points. For each closed orbit of any cubic system $E_3(a,b)$, it is cut by any straight line in at most four points [54].

Concerning a lower bound and the relative positions of limit cycles of cubic systems, some results have been obtained. Sibirskii [24] and Shi [25] gave examples of E_3 with five concentric limit cycles by perturbing a fine focus. Recently, Blows and Lloyd [19] also constructed a class of such system E_3 with five cycles. All the cubic systems given by those authors have no quadratic terms in x and y. It follows from these examples that $H_3 \geqq 5$. From a viewpoint of relative position of limit cycles, these examples only supply a nest of limit cycles enclosing a focus. Different from that distribution of limit cycles, Il'yashenko [40] also showed that $H_3 \geqq 5$. However the corresponding limit cycle may surround more than one singular point.

The author with his co-authors [27] gave a variety of distributions of limit cycles of cubic systems shown in figure 5.

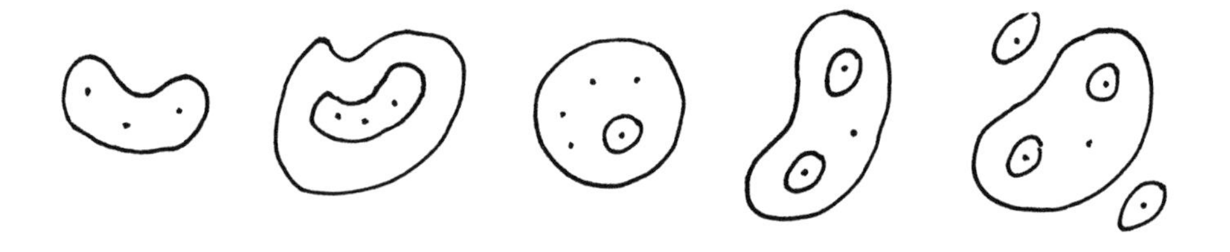

Figure 5

It should be pointed out that the famous Russian mathematician Arnold [28] obtained such a distribution of cycles of E_3 early in 1977 (figure 6).

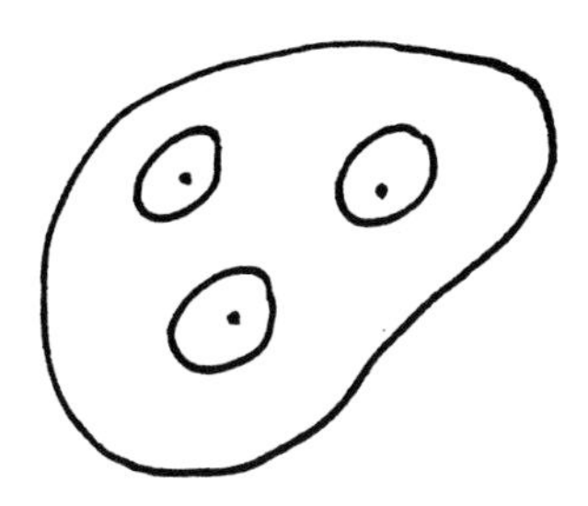

Figure 6

Li and Li [31] obtained 12 kinds of patterns of limit cycles of cubic systems by perturbing a cubic Hamiltonian system. Let C_k^m denote a kind of pattern. In C_k^m C represents limit cycle, m the number of limit cycles and k the multiplicity of singular points enclosed by the m cycles. The symbol "-" means that cycles separate. The symbol "$\supset$" means that one or several cycles enclose another or others. They obtain the following 12 kinds of patterns:

(1) C_k^1 $(k = 1,3,5,7,9)$,

(2) C_k^2 $(k = 1,3,5,7)$,

(3) $C_3^1 \supset C_1^1$,

(4) $C_3^1 \supset (C_1^1 - C_1^1)$,

(5) $C_3^1 \supset C_1^2$,

(6) $C_3^1 \supset (C_1^1 - C_1^2)$,

(7) $C_3^1 \supset C_1^2 - C_1^2)$,

(8) $C_3^2 \supset C_1^1$,

(9) $C_3^2 \supset (C_1^1 - C_1^1)$,

(10) $C_1^1 \supset C_1^1 - (C_1^1 - C_1^1)$,

(11) $C_9^1 \supset C_1^1 - (C_1^1 - C_1^1 - C_1^1)$,

(12) $C_1^1 - (C_3^1 \supset C_1^1 - C_1^1) - C_1^1$.

For example

Now let us explain the method (b) of producing limit cycles.

Before giving the newest results of Hilbert's 16th problem for E_3 we introduce the corresponding new method of producing limit cycles of E_3 which is a generalization of continuous variation of coefficients suggested by Hilbert. This method is a wonderful application of Pontryagin [29] and Zhang's [30] method to Hilbert's 16th problem for E_3. Moreover, the technique of the application is well-developed by Li and Li [31] in 1985. Li and Huang obtained that $H_3 \geqq 11$ [32]. They found a cubic system

$$\begin{aligned} \dot{x} &= x(1 + 4x^2 - y^2) + \mu y(x^2 + 0.43y^2 - \lambda) \\ \dot{y} &= y(1 + x^2 - 0.5y^2) + \mu x(x^2 + 0.43y^2 - \lambda) \end{aligned} \qquad (1)_\mu$$

where λ, μ are parameters

$$-1.579338 < \lambda < -1.57401, \quad 0 < \mu << 1.$$

System $(1)_\mu$ has at least 11 limit cycles as shown in figure 7.

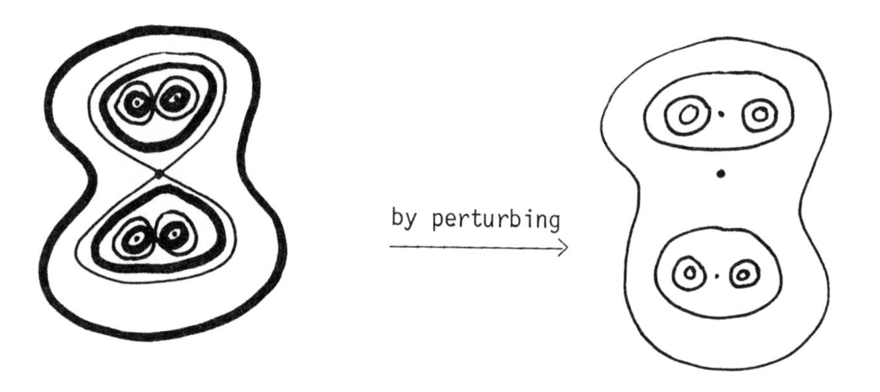

7 families of closed orbits (non-isolated)

(a) The phase portrait of $(1)_{\mu=0}$ with 4 centres and 3 saddles

(b) 11 limit cycles of (1) with $-1.579338 < \lambda < -1.57401$ for $0 < \mu <<$

Li's example with 11 cycles: $(1)_\mu$

Figure 7

Consider the disturbed Hamiltonian system:

$$\frac{dx}{dt} = \frac{\partial H}{\partial y} - \mu x[p(x,y) - \lambda] = P(x,y,\mu,\lambda)$$

$$\frac{dy}{dt} = -\frac{\partial H}{\partial x} - \mu y[q(x,y) - \lambda] = Q(x,y,\mu,\lambda) \qquad (2)_\mu$$

where $H(x,y)$, $p(x,y)$, $q(x,y)$ are polynomials with $p(0,0) = q(0,0) = 0$, λ, μ two parameters with $0 < \mu \ll 1$ and $P(0,0,\mu,\lambda)=Q(0,0,\mu,\lambda) = 0$. Suppose that the following condition holds.

CONDITION A: The integral of the system $(2)_{\mu=0}$ $H(x,y) = h$ $(h_1 < h < h_2)$ represents a family of closed orbits $\{\Gamma^h\}$ which surround C $(C \geqq 1)$ centres and/or s $(s \geqq 0)$ saddles, expand when h increases, and are negatively oriented. Let

$$f(x,y) = xp'_x + yq'_y + p + q, \quad D^h = \text{Interior of } \Gamma^h$$

$$\phi(h) = \iint_{D^h} dx\, dy, \quad \psi(h) = \iint_{D^h} f(x,y)dx\, dy \quad (h_1 < h < h_2)$$

$$\lambda(h) = \frac{\psi(h)}{2\phi(h)} \quad (h_1 < h < h_2).$$

Then the function $\lambda = \lambda(h)$ is called the detection function of $(2)_\mu$ corresponding to the family $\{\Gamma^h\}$ and the graph of this function is the detection curve.

Applying Pontryagin and Zhang's method to system $(2)_\mu$ leads to:

FUNDAMENTAL THEOREM A (on bifurcations from periodic cycles): For a fixed μ $(0 < \mu \ll 1)$ and a given value λ_0 of parameter λ, the system $(2)_\mu$ is considered. Suppose that condition A holds. Then

(i) system $(2)_\mu$ with $\lambda = \lambda_0$ has a unique stable (unstable) limit cycle near the closed curve Γ^{h_i}, if $\lambda(h_i) = \lambda_0$, $\lambda'(h_i) > 0$ (< 0),

(ii) system $(2)_\mu$ with $\lambda = \lambda_0$ has no more than k limit cycles near $\Gamma^{h_i^*}$ $(i \geqq 1)$, if $\lambda_0 = \lambda(h_i^*)$ and $\lambda'(h_i^*) = \ldots = \lambda^{(k-1)}(h_i^*) = 0$, $\lambda^{(k)}(h_i^*) \neq 0$.

If the closed orbits $\{\Gamma^h\}$ in condition A expand when h decreases or are positively oriented then the limit cycle mentioned in (i) of the theorem has the opposite stability.

CONDITION B: (1) Condition A holds. (2) Integral curves $H(x,y) = h_i$, $(i = 1,2)$ are homoclinic orbits or heteroclinic orbits of $(2)_{\mu=0}$ which surround some orbits of $\{\Gamma^h\}$ and are surrounded by some orbits of $\{\Gamma^h\}$ being symmetrical with respect to the x-axis. (3) Γ^h changes monotonically with h in each region divided by the homoclinic and heteroclinic orbits. (4) The vector field of $(2)_\mu$ is invariant under a rotation through $2\pi/k$ $(k = 1,2,\ldots)$. (5) Saddles of $(2)_\mu$ are hyperbolic. (6) $\lambda(h) \neq 0$ $(h_1 < h < h_2)$.

Applying Mel'nikov's [33,34] method to $(2)_\mu$ leads to

FUNDAMENTAL THEOREM B (on bifurcations from singular closed orbits): Suppose that conditions A, B hold. If $\lambda = \lambda(h_2) + o(\mu)$ or $\lambda = \lambda(h_1) + o(\mu)$, then $(2)_\mu$ has homoclinic or heteroclinic orbits.

By using the two fundamental theorems on bifurcations and starting with a suitably chosen Hamiltonian polynomial system, a variety of distributions of limit cycles may be obtained by perturbing the system. This is the theoretical base of the above-mentioned example with 11 limit cycles.

Concerning H_n Otpokov [35] obtained in 1954 an estimate

$$H_n \geq \begin{cases} \frac{1}{2}(n^2 + 5n) - 7, & \text{if } n \text{ is even and } n \geq 6 \\ \frac{1}{2}(n^2 + 5n) - 13, & \text{if } n \text{ is odd and } n \geq 9. \end{cases}$$

Shen [36] recently gave by means of many properties of Chebyshev's polynomials $T_n(x)$ the following recursive inequality on H_n

$$H_n \geq \left(\frac{n+1}{p}\right)^2 H_{p-1}, \text{ if } \frac{n+1}{p} \text{ is an integer.}$$

Thus

$$H_5 \geqq (\frac{6}{3})^2 H_2 \geqq 4 \times 4 = 16$$

$$H_7 \geqq (\frac{8}{4})^2 H_3 = 4H_3 \geqq 4 \times 11 = 44$$

$$H_{11} \geqq (\frac{12}{4})^2 H_3 \geqq 9 \times 11 = 99$$

...

$$H_n \geqq (\frac{n+1}{4})^2 H_3 \geqq \frac{11}{16}(n+1)^2, \text{ if } n+1 \text{ is divisible by } 4.$$

On the other hand, by Otpokov's estimation, we have

$$H_{11} \geqq 75$$

...

Therefore, Shen's result is better than Otpokov's in the case of $n = 4k - 1$ (k is a natural number).

ACKNOWLEDGEMENT: This research was supported by the China Natural Science Foundation Grant No. 1860497.

REFERENCES

[1] Hilbert, D., Mathematische Probleme, Gesammelte Abhandlungen, B.III, S.317, 1900.

[2] Aleksandrov, P.S., Problemy Gil'berta, Izdat Nauka, Moscow, 1969 (German transl.: Hilbertische Probleme, Akademie Verlag, Leipzig, 1971).

[3] Browder, F.E., Mathematical Developments Arising from Hilbert Problems, Vols. I, II, Proc. Symp. Pure Math., Vol. 28, 1976.

[4] Dulac, M.H., Sur les cycles limites, Bull. Soc. Math. Fr., 51 (1923), 45-188.

[5] Il'yashenko, Ju S., Limit cycles of polynomial vector fields with non degenerate singular points in the plane (in Russian), Funct. Anal. Appl., 18 (1984), No. 3, 32-42.

[6] Chicone, C. and Schafer, D., Separatrix and limit cycles of quadratic systems and Dulac's theorem, Trans. Amer. Math. Soc., 278 (1983), No. 2, 585-612.

[7] Bamon, R., Solution of Dulac's problem for quadratic vector fields, An. Acad. Brasil, Ciênc., 57 (3) (1985), 265-266.

[8] Bautin, N.N., On the number of limit cycles which appear with the variation of coefficients from an equilibrium position of focus or center type (in Russian), Amer. Math. Soc. Transl. No. 100, 1954.

[9] Tung, C.C., Positions of limit cycles of the systems

$$\frac{dx}{dt} = \sum_{i+j=0}^{2} a_{ij}x^i y^j, \quad \frac{dy}{dt} = \sum_{i+j=0}^{2} b_{ij}x^i y^j.$$

Sci. Sinica, 8 (1956), 151-171.

[10] Petrovskii, I.G. and Landis, E.M., On the number of limit cycles of the equation dy/dx = P(x,y)/Q(x,y), where P and Q are polynomials of the second degree, Mat. Sb. N.S. 37 (79) (1955), 209-250 (in Russian); Amer. Math. Soc. Transl. (2), 16 (1958), 177-211.

[11] Petrovskii, I.G. and Landis, E.M., On the number of limit cycles of the equation dy/dx = P(x,y)/Q(x,y), where P and Q are polynomials, Mat. Sb. N.S. 43 (85) (1957), 149-168 (in Russian); Amer. Math. Soc. Transl. (2), 14 (1960), 181-200.

[12] Landis, E.M. and Petrovskii, L.G., Letter to the Editor, Mat. Sb., 73 (115) (1967), 160; English transl.: Math. USSR, 2, (1967), 144.

[13] Ye, Y.Q., Theory of Limit Cycles (in Chinese), Shanghai Sci. Tech. Press, 1965.

[14] Shi, S.L., A concrete example of the existence of four limit cycles for plane quadratic systems, Sci. Sinica (Engl. edition) 23 (1980), 153-158; Sci. Sinica (Chinese edition) 11, (1979), 1051-1056.

[15] Chen, L.S. and Wang, M.S., The relative position and number of limit cycles of quadratic differential systems, Acta. Math. Sinica, 22 (1979), 751-758.

[16] Qin Yuanxun, Cai Suilin and Shi Songling, On limit cycles of planar quadratic system, Sci. Sinica Ser. A. 25 (1982), 41-50.

[17] Li, C.Z., Two problems of planar quadratic systems, Sci. Sinica, Ser. A, 26 (1983), 471-481.

[18] Wanner, G., Jahrbuch Uberblicze Mathematik (1983), 9-24.

[19] Lloyd, N.G. and Blows, T.R., The number of limit cycles centain polynomial differential equations, Proc. R. Soc. Edinb., 98A (1984), 215-239.

[20] Li, C.Z. Nonexistence of limit cycles around a weak focus of order 3 for any quadratic system, Chin. Ann. Math., 7B (2) (1986), 174-190.

[21] Ye, Y.Q., On the impossibility of (2,2) distribution of limit cycles of any real quadratic differential system, J. Nanjing Univ., 2 (1985), 161-182.

[22] Qin, Y.X., On surfaces defined by ordinary differential equations - a new approach to Hilbert's 16th problem (in Chinese), J. Northwest Univ., 1 (1984), 1-15.

[23] Cao, Y.L., On the mistakes of the paper by Qin, J. Nanjing Univ., 3 (1984), 415-423.

[24] Sibirskii, K.S., The number of limit cycles in the neighbourhood of a singular point, Differential'nye Uravnenija, 1 (1965), 53-66.

[25] Shi, S.L., Example of limit cycles for cubic systems, Acta. Math. Sinica, 18 (1975), No. 4, 300-304 (in Chinese).

[26] Tian, J.H., On general properties of cubic systems, Int. J. Math. Educ. Sci. Technol., 14, No. 5 (1983), 643-648.

[27] Li, J.B., Tian, J.H. and Xu, S.L., A survey of cubic systems (in Chinese), J. Sichuan Teachers Univ., 4 (1983), 32-48.

[28] Arnol'd, V.I., Loss of stability of self-oscillations close to resonances and versal deformations of equivalent vector fields, Funct. Anal. Appl., 11 (2) (1977), 1-10.

[29] Pontryagin, L.S., Uber Autoschwingunssyteme, die den Hamiltonschen Nahe Liegen, Phys. Sowjetunion, Band 6, Heft 1-2, (1934), 883-889.

[30] Zhang, Z.F., Dokl. Akad. Nauk SSSR, 119 (1958), 659-662.

[31] Li, J.B. and Li, C.F., Global bifurcations of planar disturbed Hamiltonian systems and distributions of limit cycles of cubic systems (in Chinese), Acta. Math. Sinica, 28, No. 4 (1985), 509-521.

[32] Li, J.B. and Huang, Q.M., Bifurcations of limit cycle forming compound eyes in the cubic system (Hilbert number $H_3 \geq 11$) (in Chinese), J. Yunnan Univ., 1 (1985), 7-16.

[33] Mel'nikov, V.K., On the stability of the center for time periodic perturbations, Moscow Math. Soc., 12 (1978), 1-57.

[34] Guckenheimer, J. and Holmes, P.J., Nonlinear Oscillations, Dynamical Systems and Bifurcation of Vector Fields, Springer-Verlag, Berlin, 1983.

[35] Otpokov, N.F., Mat. Sb., 34 (76) (1954), 127-144.

[36] Shen, Z.M., On the number of limit cycles of higher degree polynomial systems, J. Graduate School (Beijing), 3 (1986), No. 1, 1-4.

[37] Joyal, P., Thesis, Université de Montreal, 1985.

[38] Il'yashenko, Yu.S., On the finiteness problem of the number of limit cycles of polynomial vector fields in the plane, Usp. Mat. Nauk, 37, No. 4 (1982), 127.

[39] Il'yashenko, Yu.S., Singular points and limit cycles of differential equations in the real and complex plane (in Russian), Pushchino, Preprint, NIVTS, Akad. Nauk SSSR, Nat. Inst. Comput. Center, 1982.

[40] Il'yashenko, Yu.S., Occurrence of limit cycles of the equation $dw/dz = - R_z/R_w$ with polynomial R(z,w) under perturbation, Mat. Sb., 78, No. 3 (1969), 360-373.

[41] Sotomayor, J. and Paterlini, R., Quadratic vector fields with finitely many periodic orbits, Lect. Notes in Math., No. 1007, 1983, 783-766 or In. Symp. on Dynamical systems, IMPA, 1983.

[42] Il'yashenko, Yu.S., Dulac's memoir "On limit cycles" and related questions of the local theory of differential equations (in Russian), Usp. Mat. Nauk, 40 (1985), No.6 (246), 41-78, 199.

[43] Andronova, E.A., On the topology of quadratic systems with four (or more) limit cycles (in Russian), Usp. Mat. Nauk, 41 (1986), No. 2 (248), 183-184; Quadratic systems that are close to conservative with four limit cycles (in Russian), Methods of the Qualitative Theory of Differential Equations, Gor'kov. Gos. Univ., Gorki, 1983, 118-126, 166-167.

[44] Qin, Y., On surfaces defined by ordinary differential equations: a new approach to Hilbert's 16th problem, In Ordinary and Partial Differential Equations, Lecture Notes in Mathematics, Vol. 1151 (Eds. B.D. Sleeman and R.J. Jarvis). Springer-Verlag, Berlin, 1985, pp. 115-131.

[45] Cherkas, L.A. and Gayko, V.A., Bifurcations of limit cycles of a quadratic system with two parameters rotating a field (in Russian), Dokl. Akad. Nauk BSSR, 29 (1985), No. 8, 694-696, 764.

[46] Rudenok, A.E., Limit cycles of a two-dimensional autonomous system with quadratic nonlinarities (in Russian), Differential'nye Uravneniya, 21 (1985), No. 12, 2072-2082, 2203-2204.

[47] Shi, S.L. A method of constructing cycles with contact around a weak focus, J. Differential Equns, 41 (1981), 301-312.

[48] Qin, Y., Shi, S. and Cai, S., On limit cycles of planar quadratic systems, Sci. Sinica (Ser. A), 25 (1982), 41-50.

[49] Sleeman, B.D., The number of limit cycles of polynomial autonomous systems in the plane and Hilbert's 16th problem, Applied Analysis Report, University of Dundee, Report AA/862, March 1986, 1-88.

[50] Chow, S.N. and Mallet-Paret, J., Integral averaging and bifurcation, J. Differential Eqns, 29 (1977), 112-159.

[51] MACSYMA 304, Math. Lab. Group of the Laboratory for Computer Science, MIT, 1983.

[52] Fraser, C., Algebraic manipulation by computer - a sign of things to come? Bull. Inst. Math. Appl., 21 (1985), 167-168.

[53] Tian, J.H., On limit cycles of center-symmetrical quadratic systems, J. Sichuan Univ., No. 3 (1983), 9-14.

[54] Tian, J.H., On the shape of closed orbits of polynomial differential systems, Int. J. Math. Sci. Tech., 18, No. 4 (19) 617-621.

[55] Ecalle,J, Martinet, J, Robert Moussu and Ramis, J.-P., Non-accumulation des cycles-limites (I), C.R. Acad. Sci. Paris Ser. I Math. 304 (1987), 375-377.

[56] Ecalle, J, Martinet, J, Robert Moussu and Ramis, J.-P., Non-accumulation des cycles-limites (II), C.R. Acad. Sci. Paris Ser. I Math. 304 (1987), 431-434.

[57] Rychkov. G.S., The maximum number of limit cycles of the system $\dot{y} = x, \dot{x} = y - \sum_{i=0}^{2} a_i x^{2i+1}$ is two, Diff. Uravn. 11 (1973), 380-391.

Tian Jinghuang
Chinese Academy of Sciences,
Chengdu,
China

J.F. TOLAND

A homotopy invariant for dynamical systems with a first integral

1. INTRODUCTION

This lecture concerns joint work with Norman Dancer on a degree theory for T-periodic orbits of flows with a first integral where T is fixed *a priori*.

The familiar Brouwer degree may be regarded as an algebraic count of the number of solutions in an open bounded subset $\Omega \subset \mathbb{R}^n$ of the equation $f(x) = p$ when f is continuous on $\bar{\Omega}$ (the closure of Ω) and $f(x) \neq p$ when $x \in \partial\Omega$ (the boundary of Ω). It is an integer-valued homotopy invariant. In 1967 Fuller defined a rational-valued degree for dynamical systems which might be regarded as giving an algebraic count of the number of periodic orbits in Ω, irrespective of their period, of the equation $\dot{x} = f(x)$ when f has no zeros in $\bar{\Omega}$ and no periodic orbit in $\bar{\Omega}$ intersects $\partial\Omega$ (see Fuller [4], and Chow and Mallet-Paret [1].) In their respective contexts each of these is a homotopy invariant and has a natural set of applications in bifurcation theory using the methods of Rabinowitz [7]. In particular, each is invariant to small perturbations in f. Therefore if $\dot{x} = f(x) + \varepsilon g(x)$ has no periodic orbits in Ω for all ε sufficiently small, then the Fuller degree for $\dot{x} = f(x)$ in Ω, if it is defined, is necessarily zero. A special case when the Fuller degree is always zero arises if there exists a nondegenerate first integral for the equation $\dot{x} = f(x)$, i.e. when $\langle \nabla V(x), f(x) \rangle = 0$ for some real-valued functional V. Then V is constant on trajectories of $\dot{x} = f(x)$ and $\dot{x} = f(x) + \varepsilon \nabla V(x)$ has no periodic solutions in Ω for $\varepsilon \neq 0$. A similar difficulty was encountered by Dancer [2] when he considered the Brouwer degree for S^1-invariant functions on $\mathbb{R}^n$; in this case the Brouwer degree is always zero also. He coped by restricting attention to equivariant mappings which are, in addition, gradients of functionals on $\mathbb{R}^n$. His approach leans somewhat on the methods of Fuller.

Our purpose in this lecture is to show how to define a degree for all Lipschitz continuous, nonvanishing vector fields defined on $\bar{\Omega} \subset \mathbb{R}^n$ (where Ω is bounded and open) which are orthogonal to ∇V, V being a continuously differentiable function on Ω with nonvanishing gradient, such that every T-periodic orbit of $\dot{x} = f(x)$ in $\bar{\Omega}$ lies in Ω. If these conditions are satisfied we say that (Ω,f,V) is *admissible*. (We require f to be defined on $\bar{\Omega}$ and V to be defined on Ω and $\langle f(x),\nabla V(x)\rangle = 0$ on Ω.) Here $T > 0$ is specified *a priori*, and is fixed.

Our degree function, which will be denoted by $\deg(\Omega,f,V)$, where its dependence on T is understood but suppressed for the sake of a convenient notation, can be regarded as an algebraic count of the number of T-periodic orbits of $\dot{x} = f(x)$ in Ω. It is a homotopy invariant in the sense made precise in (II) below.

The approach described here is naïve. The degree is first defined in a smooth nondegenerate situation, and then extended to the general context by approximation. To this end one needs to establish a type of Kupka-Smale genericity result (see e.g. Palis and de Melo [6]) in the class of vector fields with a first integral. This step requires greatest technical effort and an outline of the details involved is given in section 3. In [3] there is a proof that, when the degree is calculated for a Hamiltonian system whose first integral is the Hamiltonian, then the degree agrees (up to a change of sign) with the abstract degree defined by Dancer [2] for S^1-invariant gradient mappings once the latter has been calculated for a (gradient) Hamiltonian system. We ignore this aspect of the theory now. Our presentation means that computations are carried out directly with the dynamical system, and not through an abstract formulation as in the case for the S^1-degree. We return to the observed connection with the S^1-invariant gradient degree in the later paper when we consider changes of the index and bifurcation theory [3].

This lecture is completely descriptive and deliberately avoids a detailed discussion of the technical mathematical treatment which is being reported upon. Our aim is to give an intuitive account of the degree theory beginning in section 2 with the definition and basic properties. In section 3 there is a nonrigorous descriptive account of the ideas involved in establishing the generic results on dynamical systems with a first integral on which the definitions and proofs of section 2 are based.

2. THE DEGREE

Suppose that (Ω,f,V) is admissible and that γ is the only T-periodic orbit in Ω of $\dot{x} = f(x)$ and let $p \in \gamma$. There is no loss of generality in supposing that $p = 0$, the origin of $\mathbb{R}^n$. Then 0 is a fixed point of the time-T-map F defined by the flow $\dot{x} = f(x)$, and indeed 0 is locally a unique fixed point of

F in the hyperplane $f(0)^{\perp}$(= H, say). Let $K = \nabla V(0)^{\perp}$, and let Q be the orthogonal projection in $\mathbb{R}^n$ onto $\nabla V(0)^{\perp}$ and parallel to $\nabla V(0)$ (see figure 1).

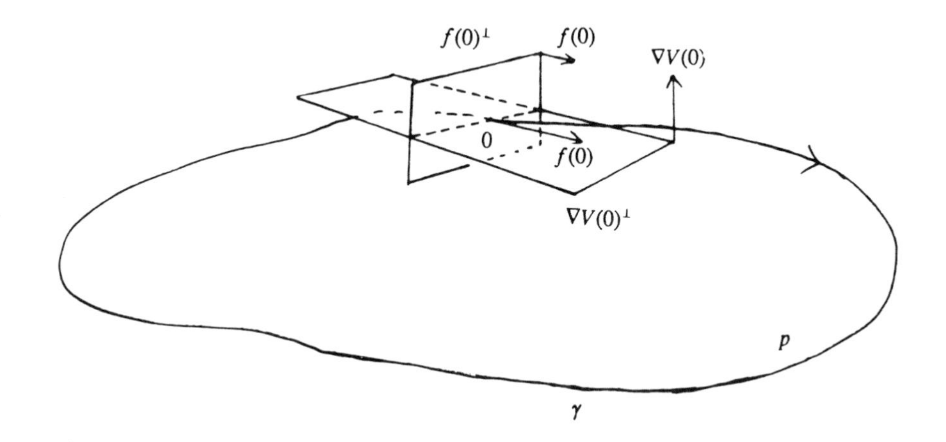

Figure 1

Now because $V(F(x)) = V(x)$ for all x and $\nabla V(0) \neq 0$, it is easy to see that there exists a neighbourhood U of 0 such that $F(x) = x$, $x \in U$, if and only if $Q(F(x)-x) = 0$. Now $\Phi = Q\circ(I-F)$ is function from an open set $U \cap H$ in the (n-1)-dimensional space H into the (n-1)-dimensional space K which has an isolated zero at $0 \in H$. Let $e_1,\ldots,e_{n-2}$ be an arbitrary basis for $H \cap K$ and let $\{\nabla V(0),e_1,\ldots,e_{n-2}\}$ and $\{f(0),e_1,\ldots,e_{n-2}\}$ be bases for H and K respectively. Then with respect to these orientations the Brouwer degree $\deg_B(U \cap H,\Phi,0)$ is defined and is independent of $\{e_1,\ldots,e_{n-2}\}$ since these basis elements are common to the domain and the co-domain. The usual stability of the Brouwer degree means that the value of the degree is locally independent of the choice of the point p on γ chosen for the calculation. Since γ is connected the calculation is independent of the choice of p on γ. Hence $\deg_B(U \cap H,\Phi,0)$ depends only on γ,f and V. If it is defined and non-zero then $\Phi(x) = 0$ has a solution in $U \cap H$ and so there is a T-periodic orbit through a point of $U \cap H$. However the following illustrates that, as it stands, this Brouwer degree lacks the stability required of an index for γ (see figure 2).

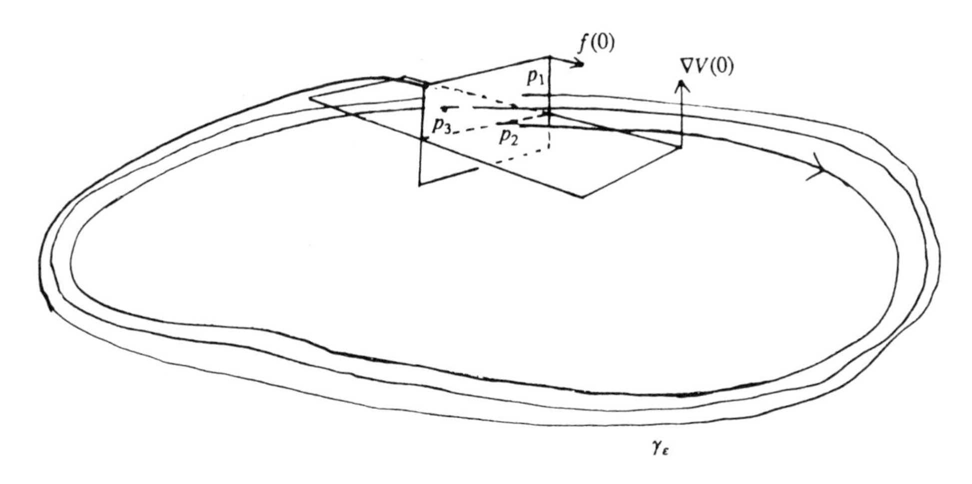

Figure 2

In the example of figure 1 let us suppose that γ has minimal period T/3, say, and that the calculation above gives $\deg_B(U \cap H,\Phi,0) = k \neq 0$. Let us suppose also that after a small perturbation of f to f_ε the T-periodic orbit γ (which must persist in some form or other because $k \neq 0$) becomes an orbit γ_ε of minimal period T. Thus Φ has three zeros, p_i, i = 1,2,3, in $U \cap H$, and if U_i is an isolating neighbourhood of p_i then $\ell = \deg(H \cap U_i,\Phi_\varepsilon,0)$ is independent of i, since p_1,p_2 and p_3 all lie on the same orbit γ_ε. Therefore, by the stability of the usual Brouwer degree, $\deg(U \cap H,\Phi,0)$, to small perturbations we find in this case that $3\ell = k$. Careful observation of what can happen in circumstances like these leads naturally to the following definition of an index for γ, an isolated T-periodic orbit. Because it is stable to perturbations, this index leads naturally to a degree theory which enjoys the usual properties.

DEFINITION: Suppose that (Ω,f,V) is admissible, that γ is the unique T-periodic solution of $\dot{x} = f(x)$ in Ω, and that T/m is the minimal period of γ. (Such an $m \in \mathbb{N}$ exists since $f(x) \neq 0$ in $\bar{\Omega}$.) Let

$$\mathrm{ind}(\gamma) = \frac{\deg_B(U \cap H,\Phi,0)}{m}.$$

If there is only a finite set $\{\gamma_1,\ldots,\gamma_k\}$ of T-periodic orbits of $\dot{x} = f(x)$ in Ω, let

$$\deg(\Omega,f,V) = \sum_{i=1}^{k} \text{ind}(\gamma_i).$$

We will say that an admissible (Ω,f,V) is *finite* if there are only a finite number of T-periodic orbits of $\dot{x} = f(x)$ in Ω. In the next section there is a description (omitting details) of the proof that for any admissible (Ω,f,V) there exists a sequence of finite admissible (Ω,f^k,V) such that $f^k \to f$ uniformly in Ω, and $\deg(\Omega,f^k,V)$ is independent of k for all k sufficiently large. This enables us to make a definition of the degree for any admissible (Ω,f,V):

$$\deg(\Omega,f,V) = \deg(\Omega,f^k,V) \text{ for all } k \text{ sufficiently large.}$$

If we show that the right-hand side is well-defined, then the following properties of the degree are immediate from the definition:

$$\text{if } \deg(\Omega,f,V) = M/N,\ (M,N) = 1, \text{ and } p^\alpha | N \text{ where } p \text{ is a prime number and } \alpha \in \mathbf{N}, \text{ then } \Omega \text{ contains an orbit of period } T/p^\alpha. \tag{I}$$

To see this, note that for all k sufficiently large the finite admissible (Ω,f^k,V) has degree (M/N), and it is immediate from the prime factorization theorem and the definition of the degree in the finite admissible case that $\dot{x} = f^k(x)$ has a (T/p^α)-periodic orbit in Ω. Since $f^k \to f$ uniformly and f is Lipschitz, the classical continuous dependence theory (Hartman [5], Ch. 2, Th. 3.2) ensures that $\dot{x} = f(x)$ also has a (T/p^α)-periodic orbit in Ω.

Suppose now that a family of finite admissible $(\Omega,f_\lambda,V_\lambda)$, $\lambda \in [0,1]$, depends continuously on λ in the sense that the f_λ's are continuous in λ with respect to the metric of uniform convergence on $\bar{\Omega}$, and the V_λ's are continuous with respect to the metric of uniform convergence on compact subsets on Ω. This will be called a *finite admissible homotopy*. Then because there are only a finite number of T-periodic orbits of $\dot{x} = f_\lambda(x)$ in Ω for each λ, it is an easy matter to infer from the homotopy invariance of the classical Brouwer degree function that $\deg(\Omega,f_\lambda,V_\lambda)$ does not depend on λ. This rather weak version of the homotopy invariance property for finite admissible homotopies is a consequence of our definition and classical continuous dependence theory for initial-value problems in a straightforward fashion.

In the next section we will indicate how to show that if $(\Omega, f_\lambda, V_\lambda)$, $\lambda \in [0,1]$, is an *admissible homotopy* in the sense that each $(\Omega, f_\lambda, V_\lambda)$ is admissible and the dependence on λ is continuous in the sense described above, then there exists a sequence of *finite admissible homotopies* $(\Omega, f_\lambda^k, V_\lambda)$ such that $f_\lambda^k \to f_\lambda$ uniformly in $\bar{\Omega}$ and $\lambda \in [0,1]$. (For each $\lambda \in [0,1]$ and $k \in \mathbb{N}$, $(\Omega, f_\lambda^k, V_\lambda^k)$ is finite and admissible.) This approximation theorem is established through a long technical argument. However, once it has been established, we can use it to prove that the above degree function is indeed well-defined, and that it enjoys a strong homotopy invariance property. The argument for well-defindedness goes as follows.

Suppose that (Ω, f, V) is admissible and $f^k \to f$ uniformly on $\bar{\Omega}$ where (Ω, f^k, V) is finite and admissible. We can show that $\deg(\Omega, f^k, V)$ is independent of k sufficiently large as follows. Let k and ℓ be natural numbers. Then for k and ℓ sufficiently large, $(\Omega, \lambda f^k + (1-\lambda)f^\ell, V)$ is an admissible homotopy since $f^k \to f$ uniformly on $\bar{\Omega}$. Now we have claimed that any admissible homotopy can be approximated by a finite admissible homotopy and so there exists $g_\lambda^m \to \lambda f^k + (1-\lambda)f^\ell$ uniformly on $\bar{\Omega}$ and $\lambda \in [0,1]$ as $m \to \infty$ where (Ω, g_λ^m, V) is finite and admissible. Hence by the homotopy invariance of finite admissible homotopies $\deg(\Omega, g_\lambda^m, V)$ is independent of $\lambda \in [0,1]$. Now $g_0^m \to f^\ell$, (Ω, f^ℓ, V) and (Ω, g_0^m, V) are both finite and admissible, and so an elementary argument involving only the stability of the Brouwer degree leads to the conclusion that $\deg(\Omega, g_0^m, V) = \deg(\Omega, f^\ell, V)$ for all m sufficiently large; also for all m sufficiently large $\deg(\Omega, g_1^m, V) = \deg(\Omega, f^k, V)$ for the same reason. Hence for all m sufficiently large

$$\deg(\Omega, f^k, V) = \deg(\Omega, g_1^m, V) = \deg(\Omega, g_0^m, V) = \deg(\Omega, f^\ell, V).$$

This shows that if $f^k \to f$ uniformly and if (Ω, f^k, V) is finite and admissible, then $\deg(\Omega, f^k, V)$ is independent of k for all k sufficiently large.

Thus provided we can show that admissible homotopies can be approximated by finite admissible homotopies then we can show that the basic definition of the degree function given above makes sense.

It is now easy to see that the approximation of admissible homotopies by finite admissible homotopies leads to the strong homotopy property of the degree function, namely that:

$\deg(\Omega, f_\lambda, V_\lambda)$ is independent of λ when $(\Omega, f_\lambda, V_\lambda)$ is any admissible homotopy. (II)

To see this let $f_\lambda^k \to f_\lambda$ uniformly on $\bar{\Omega}$ and $\lambda \in [0,1]$ be such that $(\Omega, f_\lambda^k, V_\lambda)$ is a finite admissible homotopy. Then by definition $\deg(\Omega, f_\lambda, V_\lambda) = \deg(\Omega, f_\lambda^k, V_\lambda)$ for all k sufficiently large (k depending on λ). Now if λ_1 and λ_2 are in [0,1] we obtain

$$\deg(\Omega, f_{\lambda_1}, V_{\lambda_1}) = \deg(\Omega, f_{\lambda_1}^k, V_{\lambda_1}^k) = \deg(\Omega, f_{\lambda_2}^k, V_{\lambda_2}^k)$$

$$= \deg(\Omega, f_{\lambda_2}, V_{\lambda_2})$$

for some k, where the middle equality follows because the degree function is constant for finite admissible homotopies. Thus the homotopy property is established.

So far we have outlined the definition of our degree functions for T-periodic orbits of admissible flows with a first integral and established a powerful homotopy invariance property provided we know that admissible homotopies can be approximated by finite admissible homotopies. Now it is time to show how this result is obtained. It is to be hoped that the following outline of our method, which omits the technical details, makes the rather tedious step-by-step nature of the proof clear.

3. GENERIC THEORY OF ADMISSIBLE HOMOTOPIES

Suppose that $(\Omega, f_\lambda, V_\lambda)$, $\lambda \in [0,1]$, is an admissible homotopy, i.e.

(i) $\lambda \to f_\lambda$ is continuous with respect to uniform convergence on $\bar{\Omega}$;

(ii) $\lambda \to V_\lambda$ is continuous with respect to C^1-convergence on compact subsets of Ω;

(iii) $f_\lambda(x) \neq 0$, $x \in \bar{\Omega}$, $\lambda \in [0,1]$, $\nabla V_\lambda(x) \neq 0$, $x \in \Omega$, $\lambda \in [0,1]$;

(iv) all T-periodic orbits of $\dot{x} = f_\lambda(x)$ in $\bar{\Omega}$ lie in Ω, $\lambda \in [0,1]$.

Because of (iv) we know that there is an open set U with $\bar{U} \subset \Omega$ such that all T-periodic orbits in Ω of $\dot{x} = f_\lambda(x)$, $\lambda \in [0,1]$, lie in U. Since $\nabla V \neq 0$

on $\bar{U}$ we know that $|\nabla V| \geqq \alpha > 0$ on U for some α. Since in defining the degree Ω could clearly be replaced by U there is therefore no loss of generality in assuming throughout that

(v) $|\nabla V_\lambda(x)| \geqq \alpha > 0$ on $\bar{\Omega}$, $\lambda \in [0,1]$.

Now our purpose is to indicate how $(\Omega, f_\lambda, V_\lambda)$ can be approximated by a finite admissible homotopy. There is no loss of generality in supposing at the outset (because if necessary we can make smooth approximations and extensions) that

(vi) f_λ and V_λ are jointly infinitely differentiable with respect to $\lambda \in (-\delta,(1+\delta))$ and $x \in \mathbf{R}^n$, and that f and ∇V grow no faster than linearly at infinity; in particular for any λ, solutions of the initial-value problems $\dot{x} = f_\lambda(x)$ and $\dot{x} = \nabla V_\lambda(x)$ are unique, exist for all time and depend smoothly on the initial data and on λ.

One further elementary observation is in order:

(vii) there exists $m^* > 0$ such that any T-periodic solution in Ω of $\dot{x} = f_\lambda(x)$ has minimal period no less than T/m^*.

This is immediate from (iii).

The proof may now be organized as a sequence of steps.

STEP 1: First we indicate how to prove that the smooth admissible homotopy $(\Omega, f_\lambda, V_\lambda)$ can be approximated by a smooth admissible homotopy $(\Omega, f^k_\lambda, V_\lambda)$ such that $\dot{x} = f^k_\lambda(x)$ has only finitely many T/m^*-periodic orbits in Ω for any $\lambda \in [0,1]$. The proof depends on the following geometrical observation. For each λ, let Ω_λ denote the set of x in Ω such that $x(t) \in \Omega$ for all $t \in [0,(T/m^*)+\varepsilon)$ for some $\varepsilon > 0$ if $\dot{x} = f_\lambda(x)$ and $x(0) = x$. Clearly Ω_λ is open. Let $F_\lambda : \Omega_\lambda \to \Omega$ denote the time-(T/m^*)-map for the equation $\dot{x} = f_\lambda(x)$, and let $\mathcal{F}_\lambda$ denote its graph: thus $\mathcal{F}_\lambda = \{(x, F_\lambda(x)) : x \in \Omega_\lambda\}$. Let

$$F = \bigcup_{\lambda\in(-\delta,1+\delta)} \{\lambda\} \times F_\lambda.$$

Now for each $x \in \Omega_\lambda$, let $\theta_\lambda(x)$ denote the trajectory through x of the equation $\dot{x} = \nabla V_\lambda(x)$, and let

$$\Theta = \bigcup_{\substack{\lambda\in(-\delta,1+\delta)\\ x\in\Omega_\lambda}} \{\lambda\} \times \{x\} \times \{\theta_\lambda(x)\}.$$

Now $F \subset \mathbb{R}^{2n+1}$ is a smooth submanifold of dimension (n+1) and $\Theta \subset \mathbb{R}^{2n+1}$ is a smooth submanifold of dimension (n+2). The first observation, which is not difficult to prove, is that if $F \pitchfork \Theta$ then the intersection is a two-dimensional embedded submanifold of $\mathbb{R}^{2n+1}$ and if $(\lambda,x,y) \in F \pitchfork \Theta$, then $y = F_\lambda(x) = x$ and x lies on a (T/m^*)-periodic orbit in Ω of $\dot{x} = f_\lambda(x)$. In other words $F \pitchfork \Theta$ comprises a manifold of (T/m^*)-periodic orbits of $\dot{x} = f_\lambda(x)$. However, $F \pitchfork \Theta$ is insufficient to ensure the aim of step 1; in other words, for a fixed value of λ there may be a cylinder of (T/m^*)-periodic orbits of $\dot{x} = f_\lambda(x)$. To outlaw this possibility a further transversality condition must be satisfied.

Let A denote the vector bundle with base space

$$E = \bigcup_{\lambda\in(-\delta,1+\delta)} \{\lambda\} \times \Omega_\lambda \times R^n$$

and such that the fibre over $(\lambda,x,y) \in E$ is $L(x)$, which denotes the space of linear operators from $f_\lambda(x)^\perp$ into $\nabla V_\lambda(x)^\perp$. Then A is a smooth manifold of dimension $(n^2 + 2)$. Let $\Sigma \subset A$ be the submanifold defined by

$$\Sigma = \{(\lambda,x,y,L) : (\lambda,x,y) \in \Theta, \text{ rank } L = n-2\}.$$

The Σ has co-dimension 1 in A. Now consider the set

$$D = \{(\lambda,x,F_\lambda(x),Q_{(\lambda,x)}\circ(I-d_xF_\lambda[x])|_{f_\lambda(x)^\perp} : \lambda \in (-\delta,1+\delta), x \in \Omega_\lambda\}$$

where $Q_{(\lambda,x)}$ is the orthogonal projection onto $\nabla V_\lambda(x)^\perp$. Then $D \subset A$. If $D \pitchfork \Sigma$, then $D \cap \Sigma$ is a one-dimensional manifold. In particular, if $F \pitchfork \Theta$, and $(\lambda,x,F_\lambda(x)) \in \Theta \cap F$, but $(\lambda,x,F_\lambda(x), Q_{(\lambda,x)}\circ(I-d_xF_\lambda[x])|_{f_\lambda(x)^\perp}) \notin \Sigma$, then x lies on an isolated orbit of $\dot{x} = f_\lambda(x)$. On the other hand, if $(\lambda, x, F_\lambda(x)) \in \Theta \cap F$, and $(\lambda, x, F_\lambda(x), Q_{(\lambda,x)} \circ (I - d_x F_\lambda[x])|_{f_\lambda(x)^\perp}) \in \Sigma$, and $D \pitchfork \Sigma$ then it also follows that x lies on an isolated orbit of $\dot{x} = f_\lambda(x)$ in Ω. (This occurs at the points where $\lambda = \lambda_i$ in figure 3.)

Therefore it will suffice for our purposes to show that f_λ can be perturbed in such a way that $\Sigma \pitchfork D$ and $F \pitchfork \Theta$ are assured. This can be done locally in a neighbourhood of a point (λ, p) with $p = F_\lambda(p)$ using perturbations of the form

$$f_\lambda(x) + \phi\left(\frac{x-p}{\eta}\right) \left\{a - \frac{\langle a, \nabla V_\lambda(x)\rangle}{\|\nabla V_\lambda(x)\|^2} \nabla V_\lambda(x)\right\}$$

where $\eta > 0$, ϕ is a test function with support centred at the origin, for some fixed $a \in R^n$. The proof of this result is a straightforward calculation of the derivatives of functions defined by the differential equation, *but it involves in an essential way the fact that* (T/m^*) *is the minimal period of* γ. A standard compactness argument then yields a perturbation for which $F \pitchfork \Theta$ and $D \pitchfork \Sigma$ on Ω of the form

$$f_\lambda(x) + \sum_{i=1}^{m} \phi_i\left(\frac{x-p_i}{\eta_i}\right) \left\{a_i - \frac{\langle a_i, \nabla V_\lambda(x)\rangle}{\|\nabla V_\lambda(x)\|^2} \nabla V_\lambda(x)\right\}.$$

The set $a_1, \ldots, a_m$ can be chosen so small that the perturbation of f_λ can be made as close to zero as we like. The proof depends crucially on the fact that *the orbits of minimal period* T/m^* *for* $\lambda \in [-\frac{1}{2}\delta, 1 + \frac{1}{2}\delta]$ *form a compact subset of* $[-\frac{1}{2}\delta, 1 + \frac{1}{2}\delta] \times \Omega$.

This perturbation has enabled us to describe in some detail the structure of all the (T/m^*)-periodic orbits in Ω for $\lambda \in (-\frac{1}{2}\delta, 1 + \frac{1}{2}\delta)$. Generically (i.e. after a small perturbation) there is at most a finite set of points $\lambda_1, \ldots, \lambda_k \in (-\frac{1}{2}\delta, 1 + \frac{1}{2}\delta)$ and all (T/m^*)-periodic orbits for other values of λ lie on curves γ_λ parametrized by λ. The set $\{\lambda_1, \ldots, \lambda_k\}$ is characterized by the fact that for λ_i there exists $p_i \in \gamma_{\lambda_i}$ such that

$$(\lambda_i, p_i, F_{\lambda_i}(p_i), Q_{(\lambda_i, p_i)} \circ (I - d_x F_{\lambda_i}[p_i])|_{f_{\lambda_i}(p_i)^\perp}) \in \Sigma \cap D.$$

Since $D \pitchfork \Sigma$ and $F \pitchfork \Theta$, these are turning points of the curves γ_λ of (T/m^*)-

periodic orbits mentioned earlier (see figure 3).

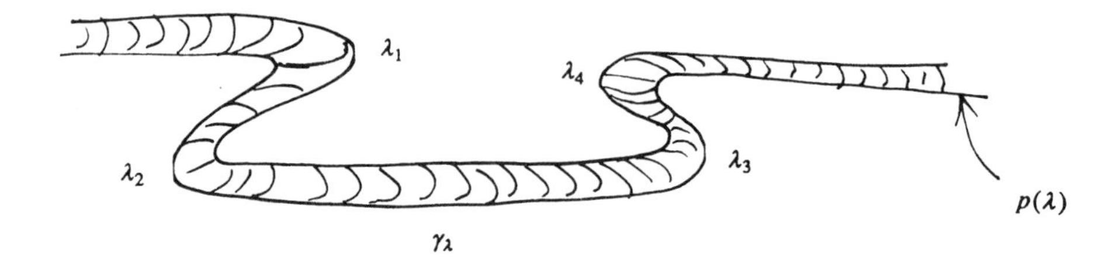

Figure 3

STEP 2: Now we make further perturbations of the vector fields f_λ for λ in small neighbourhoods of $\lambda_1,\ldots,\lambda_k$ to ensure that λ_i, $1 \leq i \leq k$, is not a period-multiplying bifurcation point. This means we must ensure that no integer root of unity, apart from unity itself, is an eigenvalue of $d_x F_{\lambda_i}[p]$ when $p = F_{\lambda_i}(p)$. To obtain this after perturbation is a matter of arguing locally in neighbourhoods of these critical (T/m^*)-periodic orbits. The perturbation is chosen in such a way that on these critical orbits the vector fields are unchanged, and hence the orbits themselves are unchanged. In a deleted neighbourhood of these orbits the vector field is adjusted slightly to ensure that no integer root of unity is an eigenvalue of $d_x F_{\lambda_i}[p_i]$, $1 \leq i \leq k$, when $F_{\lambda_i}(p_i) = p_i$. Care must be taken to ensure that after perturbation $\langle f_\lambda, \nabla V_\lambda \rangle = 0$.

STEP 3: At this stage the (T/m^*)-periodic solutions of $\dot{x} = f_\lambda(x)$ lie on curves parametrized by $\lambda \neq \lambda_i$, $1 \leq i \leq k$; the (T/m^*)-periodic solutions of $\dot{x} = f_{\lambda_i}(x)$ are isolated for each λ_i, and there is no period-multiplying bifurcation at these critical values of λ. Let us consider a curve $p(\lambda)$, $\lambda \in (\lambda_i, \lambda_{i+1})$, $p(\lambda) \in \gamma_\lambda$, where γ_λ is a (T/m^*)-periodic orbit of $\dot{x} = f_\lambda(x)$ in Ω.

This step is to show that if for some $\lambda \in (\lambda_i, \lambda_{i+1})$ period-multiplying bifurcation occurs, then after perturbation we can ensure that it is a period-doubling bifurcation which is not "vertical". Clearly, generically we cannot preclude the possibility of an eigenvalue of $d_x F_\lambda[p_\lambda]$ passing through -1 as λ varies. However it is intuitively obvious, and it can be proved, that after a perturbation no root of unity apart from 1 or -1 is an eigenvalue of $d_x F_\lambda[p_\lambda]$. Moreover the bifurcation equation for solutions of the equation $x = F_\lambda(x)$ corresponding to period-doubling is a nondegenerate cubic in one variable, and so the period-doubling bifurcation is a pitchfork. Because of translation invariance of the differential equation this corresponds to a "nonvertical" bifurcation of a single branch of $(2T/m^*)$-periodic solutions of the flow (see figure 4).

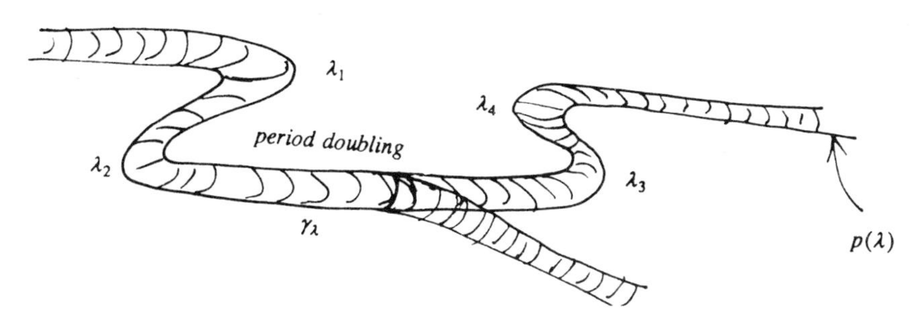

Figure 4

In particular these bifurcating solutions of period $(2T/m^*)$ lie on a curve which locally intersects $\lambda = \lambda_i$ once at the bifurcation point, and which lies locally on one side of λ_i or the other. (The quadratic term in the bifurcation equation necessarily is zero, a fact that can be observed from the translation invariance of autonomous ordinary differential equations.)

We note also that on the bifurcating branch of $(2T/m^*)$-periodic solutions no root of unity (apart from unity itself) is an eigenvalue of $d_x F[p]$, and hence close to bifurcation there is no further period-multiplying bifurcation.

At this stage we observe that all the properties of the (T/m^*)-periodic orbits established in steps 1-3 above by suitable perturbations are stable in the sense that they will continue to hold after further perturbation, provided that the further perturbation is sufficiently small. With this observation in mind we turn our attention to periodic orbits of minimal period $T/(m^*-1)$ for $\lambda \in [-\frac{1}{4}\delta, 1 + \frac{1}{4}\delta]$. This set of orbits does *not* form a compact subset of $[-\frac{1}{4}\delta, 1 + \frac{1}{4}\delta] \times \Omega$.

<u>STEP 4</u>: However, because of step 3 we know exactly that the only possible points of its boundary which are not in the set are period-doubling points with $\lambda \notin \lambda_1 \ldots \lambda_k$. We have already seen in step 3 that for λ close to the

period-doubling points and in a neighbourhood in Ω of the period-doubling orbits there are at most a finite number of $(2T/m^*)$-periodic orbits. We can therefore restrict our attention to the compact set of orbits of minimal period $(2T/m^*)$ lying outside open neighbourhoods of the period-doubling bifurcation points of step 3 above.

Now we repeat the argument of step 1, with the time-$(T/(m^*-1))$-map instead of the time (T/m^*)-map used earlier. This then ensures that after a small perturbation there are only a finite number of orbits of minimal period $(T/(m^*-1))$ in Ω for any $\lambda \in [-\frac{1}{4}\delta, 1 + \frac{1}{4}\delta]$, and that these orbits lie on curves parametrized by λ except for a finite set of turning points. Now we repeat the argument of step 2 to ensure that the turning points are not period-doubling bifurcation points, and finally we repeat the argument of step 3 to ensure that any period-multiplying bifurcations which might arise are period-doubling bifurcations, and that the bifurcation is a nondegenerate pitchfork on which, locally, no period-multiplying bifurcation occurs.

With this in hand we turn our attention to the set of orbits of minimal period $T/(m^*-3)$.

Now we proceed by induction, and after m^* iterations we arrive at a smooth finite admissible homotopy.

Of course if (Ω, f, V) is admissible, then it can be thought of as a constant admissible homotopy, and so the approximation result needed in the definition of the degree has been obtained in passing.

REFERENCES

[1] Chow, S.N. and Mallet-Paret, J., The Fuller index and global Hopf bifurcation, J. Diff. Eqns, 29 (1978), 66-85.

[2] Dancer, N., A new degree for S^1-invariant gradient mappings and applications, Ann. Inst. H. Poincaré, Analyse non linéare, 2 (1985), 329-370.

[3] Dancer, N. and Toland, J.F., Bifurcations of periodic orbits for systems with a first integral, to appear.

[4] Fuller, F.B., An index of fixed point type for periodic orbits, Amer. J. Math., 89 (1967), 133-148.

[5] Hartman, P., Ordinary Differential Equations, 2nd Edition, Birkhäuser, Basel, 1982.

[6] Palis (Jr), J. and de Melo, W., Geometric Theory of Dynamical Systems. An Introduction, Springer-Verlag, Berlin, 1982.

[7] Rabinowitz, P.H., Some global results for nonlinear eigenvalue problems, J. Funct. Anal., 7 (1971), 487-513.

[8] Stein, E.M., Singular Integrals and Differentiability Properties of Functions, Princeton University Press, Princeton, 1970.

J.F. Toland
School of Mathematical Sciences
University of Bath
Claverton Down
Bath BA2 7AY
U.K.

J.J. TYSON

Traveling waves in excitable media

1. INTRODUCTION

Many tissues of biological origin are able to transmit signals in the form of propagating waves of chemical or electrical activity. The most familiar example is the nerve axon which conducts waves of membrane depolarization along its length [1]. Neural networks, as found in the cerebral cortex, support organized waves of electrical [2] and chemical [3] activity in two spatial dimensions. Heart muscle propagates waves of electrical activity and muscular contraction. The spatial organization of these waves in two and three spatial dimensions is related to cardiac function and dysfunction [4]. Signal transmission is also important in developmental biology where spatial and temporal coordination is essential to proper morphogenesis. A paradigm of such coordination is found in the slime mold *Dictyostelium discoideum* where traveling waves of cyclic AMP direct the process of aggregation of single-celled amoebae into a multicellular slug [5].

Traveling waves of chemical reaction are also found in nonliving systems. Waves of oxidation are observed in many chemical reactions [6], the most famous of which is the Belousov-Zhabotinskii (BZ) reaction. The BZ reaction involves the oxidation of certain carboxylic acids by bromate ions in the presence of a suitable transition-metal ion catalyst. In the early 1950s Belousov was studying this reaction as an analog of the oxidative decarboxylation of organic acids in living cells when he discovered that the reaction oscillates back and forth between oxidized and reduced states for many cycles [7]. Later Zhabotinskii, Winfree, and others discovered that the BZ reaction would also support spatial waves of oxidation which propagate through thin unstirred layers of reagent [8]. A thin layer need not be spontaneously oscillatory to support oxidation waves. Indeed, Winfree's nonoscillatory recipe [8] is particularly convenient for studying wave propagation in the BZ reaction. When carefully prepared, the medium will remain for a long time uniformly in a reduced state, but if perturbed sufficiently, a single circular wave of oxidation will propagate away from the point of perturbation until it collides with the boundary of the dish and disappears.

There are many similarities between oxidation waves in the BZ reaction and activity waves in neural, neuromuscular, and developmental biology. All these media may be spontaneously oscillatory or merely excitable. They can propagate waves in one, two, or three spatial dimensions. Wave propagation is self-regenerative, i.e. waves propagate without loss of amplitude or speed. Waves are annihilated on collision with other waves or with boundaries. Periodic traveling waves show dispersion, i.e. the speed of propagation varies with wave frequency. In two spatial dimensions two characteristic patterns of propagating waves are observed: expanding concentric circular waves ("target" patterns) and rotating spiral waves. In a given preparation, target patterns generally come in a variety of temporal periods (with wavespeed and wavelength determined by the dispersion relation), whereas spiral patterns have a unique pitch and rotation frequency. In three spatial dimensions these characteristic patterns generalize to expanding spherical waves and rotating scroll waves.

The similarities among the various examples of wave propagation in excitable media can be traced to a similarity in mathematical description. Each example can be described with reasonable fidelity by a pair of nonlinear reaction-diffusion equations

$$\frac{\partial u}{\partial t} = \varepsilon D_1 \nabla^2 u + \frac{1}{\varepsilon} f(u,v)$$
$$\frac{\partial v}{\partial t} = \varepsilon D_2 \nabla^2 v + g(u,v). \qquad (1)$$

In this system of equations u and v represent the state of the system (e.g. chemical concentrations, membrane potential, ionic conductance, enzyme activity, etc.) as functions of time and space. The functions f(u,v) and g(u,v) describe the local rate of change of u and v in the absence of spatial coupling. Typical forms of f and g are illustrated in the phase plane in figure 1. ∇^2 is the Laplacian operator in one, two, or three spatial dimensions, and D_1 and D_2 are diffusion coefficients of u and v. (Space and time have been scaled so that D_1, D_2, f and g are all order-one relative to ε.) The small parameter ε bespeaks a significant separation in time-scales for u and v, with u tending to change much more rapidly than v.

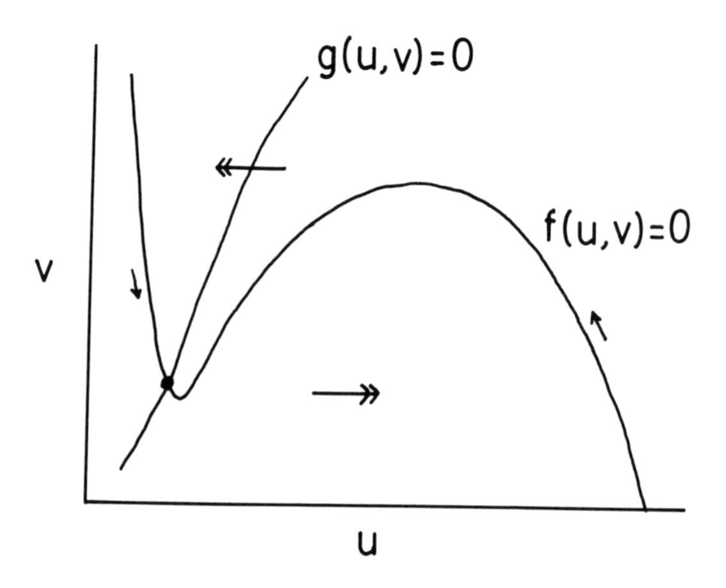

Figure 1 Phase plane illustrating the nullclines f(u,v) = 0 and g(u,v) = 0.

For the various examples of excitable media discussed so far, we can make the following associations

Medium	u	v
Belousov-Zhabotinskii reaction	Bromous acid	Ferroin
Neuromuscular tissue	Membrane potential	Ionic conductance
Dictyostelium discoideum	Cyclic AMP	Membrane receptor

Among these examples only the names of the state variables are changed and certain quantitative details of the kinetic functions f and g. Qualitative features of the solutions of system (1) carry over directly to all cases, and it is these qualitative features that we now review.

2. WAVES IN ONE AND TWO SPATIAL DIMENSIONS

Qualitative and quantitative information about traveling wave solutions to system (1) in one spatial dimension can be obtained by singular perturbation theory [9]. The basic result is the demonstration of propagating fronts that switch the system, at constant v, from the left-hand branch of $f(u,v) = 0$ to the right-hand branch of $f(u,v) = 0$. From such fronts (red → blue) one can construct isolated traveling pulses (red → blue → red) and periodic traveling waves (...red → blue → red → blue...). As mentioned, the speed of periodic traveling waves depends on period; a typical dispersion relation is illustrated in figure 2.

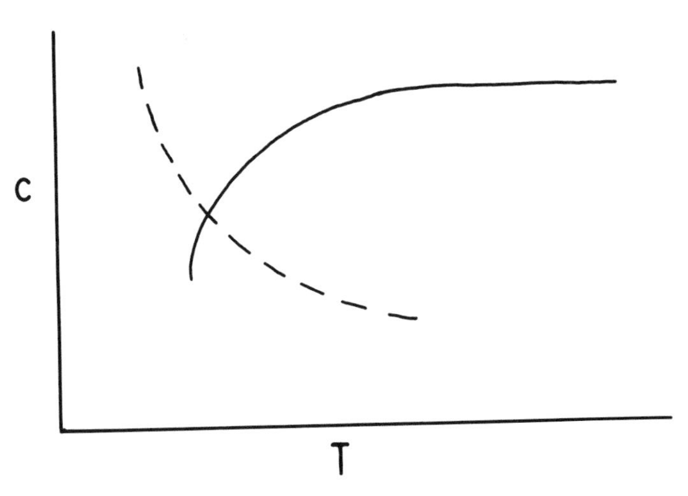

Figure 2 Dispersion relation (full curve) and curvature relation (broken curve).

The characteristic periodic patterns in two spatial dimensions, targets and spirals, must satisfy the dispersion relation because sufficiently far from the center of either pattern in a radial direction both targets and spirals are identical to one-dimensional periodic traveling waves. Target patterns need only satisfy the dispersion relation. That is, given any temporal period (T) above the minimum period in figure 2, the asymptotic speed (c) of propagation is fixed by the dispersion relation and the wavelength is simply $\lambda = cT$. Spiral waves on the other hand seem to obey another constraint in addition to the dispersion relation because in a given medium the rotation frequencies of all spiral waves are the same.

The additional constraint on spiral waves arises from consideration of the effects of wavefront curvature on speed of propagation. These effects have been uncovered in a series of papers by Zykov [10], Keener and Tyson [11], and others [12]. These authors show that the normal velocity of a wavefront (N) is equal to the speed of plane-wave propagation (c) adjusted by an amount proportional to the curvation (K) of the front:

$$N = c + \varepsilon D_1 K. \tag{2}$$

For positive curvature (wavefront curved in the direction of propagation) $N > c$, whereas for negative curvature (wavefront curved away from its direction of propagation) $N < c$. The curvature relation (2) has been derived by other authors in other contexts (crystal growth, spreading flames) [13].

To see how (2) constrains spiral waves, consider the parametric equations for a rigidly rotating one-armed spiral

$$x = r\cos[\theta(r) - \omega t]$$

$$y = r\sin[\theta(r) - \omega t].$$

Here $\theta(r)$ determines the shape of the spiral (at fixed t) and ω is the angular frequency of rotation ($\omega = 2\pi/T$). Our problem is to determine both $\theta(r)$ and ω. Since N depends on $\theta'(r)$ and ω, and K depends on $\theta'(r)$ and $\theta''(r)$, (2) is really an ordinary differential equation for the unknown function $\theta(r)$ in terms of two parameters c and ω. Applying end conditions (at $r = 0$ and $r \to \infty$, say) to this ODE, we obtain a typical eigenvalue problem which determines a unique c for each value of ω. A rough-and-ready approximation to this curvature constraint is $c = (6\pi\varepsilon D_1/T)^{1/2}$, which is plotted in figure 2. Keener and Tyson [11] have emphasized the point that spiral waves should lie at the intersection of the dispersion relation and the curvature relation, as illustrated in figure 2.

The view of curvature and spiral waves has been tested in a number of ways. Direct experimental confirmation of (2) has been obtained for oxidation waves in the BZ reaction [14]. Keener and Tyson [11] have compared their theory in detail with experimental measurements of spiral waves in BZ reagent and with numerical solutions of PDE (1) with Oregonator kinetics (a reasonable model of the BZ reaction). Furthermore, Tyson and coworkers [15] have compared the dispersion/curvature theory with numerical solutions of PDE (1) with modified FitzHugh-Nagumo kinetics (a reasonable model of heart tissue) and with Martiel-Goldbeter kinetics (a reasonable model of cyclic AMP waves in *Dictyostelium*). In all cases there is good agreement between theory and numerics, and where available between theory and experimental observations.

3. TRAVELING WAVES IN THREE SPATIAL DIMENSIONS

In three-dimensional space spiral waves become scroll-shaped waves rotating around a one-dimensional filament which threads through the spatial domain, either intersecting the boundary or closing on itself in a ring [16]. During the course of many rotations of the scroll wave around the filament, the filament itself moves through space. If we knew the laws of motion of the filament, we could predict the entire history of the three-dimensional scroll wave, so these laws become the focus of study of wave propagation in three dimensions.

The filament moves because it is pulled about by the rotating scroll wave which at any instant in any local region is attempting to move with normal velocity $N = c + \varepsilon D_1(K_1 + K_2)$, where K_1 and K_2 are the principal curvatures of the wavefront surface. Keener [17] has used this notion to derive a set of equations describing the motion of the filament. Keener's equations have the form

alteration in rotation rate of scroll wave around filament $= c_1\kappa - a_1 w^2 + b_1 \dfrac{\partial w}{\partial s}$

normal component of velocity of filament $= b_2\kappa - a_2 w^2 + c_2 \dfrac{\partial w}{\partial s}$

binormal component of velocity of filament $= c_3\kappa - a_3 w^2 + c_4 \dfrac{\partial w}{\partial s}$

where s = arc length along filament, $\kappa(s,t)$ = curvature of filament, and $w(s,t)$ = twist rate of scroll wave around the filament as measured in the laboratory frame of reference. The coefficients a_i, b_i, c_i are constants which depend on the matrix of diffusion coefficients and the form of the spiral wave solution to the two-dimensional problem. In the simple case of equal diffusion coefficients ($D_1 = D_2$), $b_1 = b_2 = D$ and $c_1 = c_2 = c_3 = c_4 = 0$. If, furthermore, the filament is untwisted and untorted ($w = 0$), then Keener's equations reduce to the simple relation $n = D\kappa$, where n is the normal velocity of the filament in its tangent plane.

The equation $n = D\kappa$ is the simplest equation of motion for a scroll wave filament. It has been derived by many people in diverse ways [18,19] and applied primarily to the case of scroll rings. If r is the radius of a circular filament, then $n = D\kappa$ implies that $dr/dt = -D/r$, or $r(t) = (r_0^2 - 2Dt)^{1/2}$. That is, scroll rings should shrink and vanish in finite time. Such behavior is observed in numerical calculations on PDE (1) [18] and in experimental observations of BZ scroll rings [20]. Keener and Tyson [19] have also applied $n = D\kappa$ to the case of elongated spiral waves and elongated target patterns observed in thick layers of BZ reagent [16], and they found remarkable agreement between theory and experiment.

The remaining challenge is to solve Keener's equations for filament motion for more complicated situations, and to compare theoretical predictions with numerical calculations on three-dimensional excitable media [21] and with yet-to-be-obtained experimental measurements of scroll wave evolution.

ACKNOLWEDGEMENT: Lengthy discussions over many years with Jim Keener, Art Winfree and Paul Fife have shaped my ideas about traveling waves in excitable media. This work was supported in part by the National Science Foundation.

REFERENCES

[1] A.L. Hodgkin and A.F. Huxley, A quantitative description of membrane current and its application to conduction and excitation in nerve, J. Physiol., 117 (1952), 500-544.

[2] H. Petsche, O. Prohaska, P. Rappelsberger, R. Vollmer and A. Kaiser, Cortical seizure patterns in multidimensional view: the information content of equipotential maps, Epilepsia, 15 (1974), 439-463.

[3] V.I. Koroleva and J. Bures, Circulation of cortical spreading depression around electrically stimulated areas and epileptic foci on the neocortex of rats, Brain Res., 173 (1979), 209-215; N.A. Gorelova and J. Bures, Spiral waves of spreading depression in the isolated chicken retina, J. Neurobiol., 14 (1983), 353-363.

[4] A.T. Winfree, When Time Breaks Down, Princeton University Press, Princeton, NJ, 1987.

[5] J.T. Bonner, Hormones in social amoebae and mammals, Sci. Amer., 220 (June 1969), 78-91.

[6] R. Luther, Raumliche Fortpflanzung Chemischer Reaktionen, Z. Elektrochem., 12 (1906), 596-600; K. Showalter, Chemical waves, in Kinetics of Nonhomogeneous Processes, G.R. Freeman, Ed., John Wiley, New York, 1987, pp. 769-821.

[7] B.P. Belousov, A periodic reaction and its mechanism, in Oscillations and Traveling Waves in Chemical Systems, R.J. Field and M. Burger, Eds., John Wiley, New York, 1985, pp. 605-613; A.T. Winfree, The prehistory of the Belousov-Zhabotinsky oscillator, J. Chem. Educ., 61 (1984), 661-663.

[8] A.N. Zaikin and A.M. Zhabotinskii, Concentration wave propagation in two-dimensional liquid-phase self-oscillating system, Nature, 225 (1970), 535-537; A.T. Winfree, Spiral waves of chemical activity, Science, 175 (1972), 634-636; J.A. DeSimone, D.L. Beil and L.E. Scriven, Ferroin-collodion membranes: dynamic concentration patterns in planar membranes, Science, 180 (1973), 946-948.

[9] P. Ortoleva and J. Ross, Theory of propagation of discontinuities in kinetic systems with multiple time scales: fronts, front multiplicity, and pulses, J. Chem. Phys., 63 (1975), 3398-3408; L.A. Ostrovskii and V.G. Yakhno, Formation of pulses in an excitable medium, Biophysics, 20 (1975), 498-503; P.C. Fife, Singular perturbation and wave front techniques in reaction-diffusion problems, SIAM-AMS Proc., 10 (1976), 23-50; J.J. Tyson and P.C. Fife, Target patterns in a realistic model of the Belousov-Zhabotinskii reaction, J. Chem. Phys., 73 (1980), 2224-2237.

[10] V.S. Zykov, Kinematics of the steady circulation in an excitable medium, Biophysics, 25 (1980), 329-333; V.S. Zykov, Analytical evaluation of the dependence of the speed of an excitation wave in a two-dimensional excitable medium on the curvature of its front, Biophysics, 25 (1980), 906-911; V.S. Zykov, Modelling of Wave Processes in Excitable Media, Manchester University Press, Manchester, 1988.

[11] J.P. Keener, A geometrical theory for spiral waves in excitable media, SIAM J. Appl. Math., 46 (1986), 1039-1056; J.P. Keener and J.J. Tyson, Spiral waves in the Belousov-Zhabotinskii reaction, Physica, 21D (1986), 307-324; J.J. Tyson and J.P. Keener, Singular perturbation theory of traveling waves in excitable media (a review), Physica, 32D (1988) 327-361.

[12] A.S. Mikhailov and V.I. Krinskii, Rotating spiral waves in excitable media: the analytical results, Physica, 9D (1983), 346-371; P.C. Fife, Propagator-controller systems and chemical patterns, in Non-Equilibrium Dynamics in Chemical Systems, C. Vidal and A. Pacault, Eds., Springer-Verlag, Berlin, 1984, pp. 76-88; E. Meron and P. Pelcé, Model for spiral wave formation in excitable media, Phys. Rev. Lett., 60 (1988), 1880-1883.

[13] W.K. Burton, N. Cabrera and F.C. Frank, The growth of crystals and the equilibrium structure of their surfaces, Phil. Trans. R. Soc. (Lond.), 243A (1951), 299-358; G.H. Markstein, Experimental and theoretical studies of flame front stability, J. Aero. Sci., 18 (1951), 199-209; M.L. Frankel and G.I. Sivashinsky, On the nonlinear thermal diffusive theory of curved flames, J. Physique (Paris), 48 (1987), 25-28.

[14] P. Foester, S.C. Müller and B. Hess, Curvature and propagation velocity of chemical waves, Science, 241 (1988), 685-687.

[15] J.J. Tyson and J.P. Keener, Spiral waves in a model of myocardium, Physica, 29D (1987), 215-222; J.J. Tyson, K.A. Alexander, V.S. Manoranjan and J.D. Murray, Spiral waves of cyclic AMP in a model of slime mold aggregation, Physica, 34D (1989), 193-207.

[16] A.T. Winfree, Scroll-shaped waves of chemical activity in three dimensions, Science, 181 (1973), 937-939; A.T. Winfree, Two kinds of wave in an oscillating chemical solution, Faraday Symp. Chem. Soc., 9 (1974), 38-46; A.T. Winfree, Rotating chemical reactions, Sci. Amer., 230 (June, 1974), 82-95.

[17] J.P. Keener, The dynamics of three dimensional scroll waves in excitable media, Physica, 31D (1988) 269-276; J.P. Keener, Knotted scroll-wave filaments in excitable media, (preprint).

[18] A.V. Panfilov and A.M. Pertsov, Vortex ring in a three-dimensional active medium described by reaction-diffusion equations, Dokl. Biophys., 274 (1984), 58-60; A.V. Panfilov, A.N. Rudenko and V.I. Krinskii, Vortex rings in 3-dimensional active media with diffusion in two components, Biofizika, 31 (1986), 850-854; L.V. Yakushevich, Vortex filament elasticity in active medium, Stud. Biophys., 100 (1984), 195-200.

[19] J.P. Keener and J.J. Tyson, The motion of untwisted untorted scroll waves in Belousov-Zhabotinsky reagent, Science, 239 (1988), 1284-1286.

[20] B.J. Welsh, J. Gomatam and A.E. Burgess, Three-dimensional chemical waves in the Belousov-Zhabotinskii reaction, Nature, 304 (1983), 611-614.

[21] A.S. Mikhailov, A.V. Panfilov and A.N. Rudenko, Twisted scroll waves in active three-dimensional media, Phys. Lett., 109A (1985), 246-250; A.V. Panfilov and A.T. Winfree, Dynamical simulations of twisted scroll rings in three-dimensional excitable media, Physica, 17D (1985), 323-330; P.J. Nandapurkar and A.T. Winfree, A computational study of twisted linked scroll waves in excitable media, Physica, 29D (1987), 69-83; A.V. Panfilov and A.N. Rudenko, Two regimes of scroll ring drift in three-dimensional active media, Physica, 28D (1987), 215-218.

John J. Tyson
Department of Biology
Virginia Polytechnic Institute and State University
Blacksburg,
Virginia 24061
U.S.A.